METHODS IN MOLECULAR BIOLOGY

Series Editor
John M. Walker
School of Life and Medical Sciences
University of Hertfordshire
Hatfield, Hertfordshire, AL10 9AB, UK

For further volumes:
http://www.springer.com/series/7651

miRNA Biogenesis

Methods and Protocols

Edited by

Ulf Andersson Vang Ørom

Department of Molecular Biology and Genetics, Aarhus University, Aarhus, Denmark

Humana Press

Editor
Ulf Andersson Vang Ørom
Department of Molecular Biology and Genetics
Aarhus University
Aarhus, Denmark

ISSN 1064-3745 ISSN 1940-6029 (electronic)
Methods in Molecular Biology
ISBN 978-1-4939-8623-1 ISBN 978-1-4939-8624-8 (eBook)
https://doi.org/10.1007/978-1-4939-8624-8

Library of Congress Control Number: 2018947780

Printed on acid-free paper

This Springer imprint is published by the registered company Springer Science+Business Media, LLC part of Springer Nature.
The registered company address is: 233 Spring Street, New York, NY 10013, U.S.A.

Preface

miRNAs are well-studied regulatory small RNAs affecting the translation and stability of numerous mRNAs. miRNAs are emerging through a complex biogenesis pathway going from primary transcripts that can be as long as 30 kb to mature miRNA of approximately 21 nts. miRNA biogenesis is a sophisticated series of events assuring a finely tuned regulation of miRNA expression and activity. Understanding miRNA biogenesis better could expand our knowledge of molecular events, contribute to extended molecular networks in gene regulation, and lead to development of new therapeutic interventions.

It is my hope that this protocol series will contribute to the further development of knowledge and application of miRNA biogenesis, both in the basic research lab and in the clinic.

Aarhus, Denmark *Ulf Andersson Vang Ørom*

Contents

Contributors

STEFAN L. AMERES • *IMBA—Institute of Molecular Biotechnology, Vienna Biocenter, Vienna, Austria*

MALIKA ARORA • *Multidisciplinary Research Unit, Indian Council of Medical Research, GGSMCH, Faridkot, Punjab, India*

PARVEEN BANSAL • *University Centre of Excellence in Research, Baba Farid University of Health Sciences, Faridkot, Punjab, India*

RENU BANSAL • *Department of Microbiology, GGSMCH, Faridkot, Punjab, India*

JAMES P. BROUGHTON • *Department of Dermatology, Stanford University School of Medicine, Stanford, CA, USA*

SUDHIR CHANDNA • *Division of Natural Radiation Response Mechanisms, Institute of Nuclear Medicine and Allied Sciences, Timarpur, Delhi, India*

TSUNG-CHENG CHANG • *Department of Molecular Biology, University of Texas Southwestern Medical Center, Dallas, TX, USA*

ROSS A. CORDINER • *Centre for mRNP Biogenesis and Metabolism, Department of Molecular Biology and Genetics, Aarhus University, Aarhus, Denmark*

ALEX S. FLYNT • *Department of Biological Sciences, University of Southern Mississippi, Hattiesburg, MS, USA*

JOSEPHINE GALIPON • *Institute for Advanced Biosciences, Keio University, Tsuruoka, Yamagata, Japan; Systems Biology Program, Graduate School of Media and Governance, Keio University, Fujisawa, Kanagawa, Japan; Department of Biological Sciences, Graduate School of Science, The University of Tokyo, Bunkyo-ku, Tokyo, Japan*

JAYA GANGOPADHYAY • *Department of Cell Biology, Harvard Medical School, Boston, MA, USA*

GEORGIOS GEORGAKILAS • *Department of Genetics, University of Pennsylvania, Philadelphia, PA, USA; Institute for Immunology, University of Pennsylvania, Philadelphia, PA, USA; Epigenetics Institute at Perelman School of Medicine, University of Pennsylvania, Philadelphia, PA, USA*

THOMAS B. HANSEN • *Department of Molecular Biology and Genetics (MBG) and Interdisciplinary Nanoscience Center (iNANO), Aarhus University, Aarhus, Denmark*

ARTEMIS G. HATZIGEORGIOU • *DIANA-Lab, Hellenic Pasteur Institute, Athens, Greece; Department of Electrical and Computer Engineering, University of Thessaly, Volos, Greece*

ALEXANDER IOCOLANO • *Department of Cell Biology, Harvard Medical School, Boston, MA, USA*

SOH ISHIGURO • *Institute for Advanced Biosciences, Keio University, Tsuruoka, Yamagata, Japan; Systems Biology Program, Graduate School of Media and Governance, Keio University, Fujisawa, Kanagawa, Japan; Department of Biological Sciences, Graduate School of Science, The University of Tokyo, Bunkyo-ku, Tokyo, Japan*

RINTARO ISHII • *Department of Computational Biology, Graduate School of Frontier Sciences, The University of Tokyo, Kashiwa-shi, Chiba-ken, Japan*

KRYSTIAN JAŻDŻEWSKI • *Warsaw Genomics INC, Warsaw, Poland; Genomic Medicine, Medical University of Warsaw, Warsaw, Poland; Centre of New Technologies, University of Warsaw, Warsaw, Poland*
MONIKA KOLANOWSKA • *Warsaw Genomics INC, Warsaw, Poland; Genomic Medicine, Medical University of Warsaw, Warsaw, Poland; Centre of New Technologies, University of Warsaw, Warsaw, Poland*
SHINJI KONDO • *Inter-University Research Institute Corporation, Research Organization of Information and Systems, Tachikawa, Tokyo, Japan*
MARTA KOTLAREK • *Warsaw Genomics INC, Warsaw, Poland; Genomic Medicine, Medical University of Warsaw, Warsaw, Poland*
ANNA KUBIAK • *Warsaw Genomics INC, Warsaw, Poland; Genomic Medicine, Medical University of Warsaw, Warsaw, Poland; Centre of New Technologies, University of Warsaw, Warsaw, Poland*
ASHISH KUMAR • *Texas Biomedical Research Institute, San Antonio, TX, USA*
ANNITA LOULOUPI • *Max Planck Institute for Molecular Genetics, Berlin, Germany; Free University of Berlin, Berlin, Germany*
SARA MACIAS • *Institute of Immunology and Infection Research, School of Biological Sciences, University of Edinburgh, Edinburgh, UK*
MATTEO J. MARZI • *Center for Genomic Science of IIT@SEMM, Istituto Italiano di Tecnologia (IIT), Milan, Italy*
GUNTER MEISTER • *Laboratory for RNA Biology, Biochemistry Center Regensburg (BZR), University of Regensburg, Regensburg, Germany*
JOSHUA T. MENDELL • *Department of Molecular Biology, University of Texas Southwestern Medical Center, Dallas, TX, USA; Hamon Center for Regenerative Science and Medicine, University of Texas Southwestern Medical Center, Dallas, TX, USA; Simmons Comprehensive Cancer Center, University of Texas Southwestern Medical Center, Dallas, TX, USA; Howard Hughes Medical Institute, University of Texas Southwestern Medical Center, Dallas, TX, USA*
FRANCESCO NICASSIO • *Center for Genomic Science of IIT@SEMM, Istituto Italiano di Tecnologia (IIT), Milan, Italy*
MARIKO OKADA-HATAKEYAMA • *Laboratory for Integrated Cellular Systems, RIKEN Center for Integrative Medical Sciences (IMS), Yokohama, Kanagawa, Japan; Laboratory of Cell Systems, Institute for Protein Research, Osaka University, Suita-shi, Osaka-fu, Japan*
ULF ANDERSSON VANG ØROM • *Department of Molecular Biology and Genetics, Aarhus University, Aarhus, Denmark*
AMY E. PASQUINELLI • *Division of Biology, University of California, San Diego, La Jolla, CA, USA*
NIKOS PERDIKOPANIS • *DIANA-Lab, Hellenic Pasteur Institute, Athens, Greece; Department of Electrical and Computer Engineering, University of Thessaly, Volos, Greece*
ROBIN REED • *Department of Cell Biology, Harvard Medical School, Boston, MA, USA*
MADALENA M. REIMÃO-PINTO • *IMBA—Institute of Molecular Biotechnology, Vienna Biocenter, Vienna, Austria*
ANGELA M. RODRIGUES-VIANA • *IMBA—Institute of Molecular Biotechnology, Vienna Biocenter, Vienna, Austria*

YUTAKA SUZUKI • *Department of Computational Biology, Graduate School of Frontier Sciences, The University of Tokyo, Kashiwa-shi, Chiba-ken, Japan*
MASARU TOMITA • *Institute for Advanced Biosciences, Keio University, Tsuruoka, Yamagata, Japan; Systems Biology Program, Graduate School of Media and Governance, Keio University, Fujisawa, Kanagawa, Japan*
NORA TREIBER • *Laboratory for RNA Biology, Biochemistry Center Regensburg (BZR), University of Regensburg, Regensburg, Germany*
THOMAS TREIBER • *Laboratory for RNA Biology, Biochemistry Center Regensburg (BZR), University of Regensburg, Regensburg, Germany*
KUMIKO UI-TEI • *Department of Biological Sciences, Graduate School of Science, The University of Tokyo, Bunkyo-ku, Tokyo, Japan; Department of Computational Biology, Graduate School of Frontier Sciences, The University of Tokyo, Kashiwa-shi, Chiba-ken, Japan*
ANNA WÓJCICKA • *Warsaw Genomics INC, Warsaw, Poland; Genomic Medicine, Medical University of Warsaw, Warsaw, Poland*
SHANYE YIN • *Department of Cell Biology, Harvard Medical School, Boston, MA, USA*
YONG YU • *Department of Cell Biology, Harvard Medical School, Boston, MA, USA*
MOHAMMAD FARID ZIA • *Department of Biological Sciences, University of Southern Mississippi, Hattiesburg, MS, USA*

Chapter 1

High-Throughput Characterization of Primary microRNA Transcripts

Tsung-Cheng Chang and Joshua T. Mendell

Abstract

Proper control of microRNA (miRNA) expression is critical for normal development and physiology, while abnormal miRNA expression is a common feature of many diseases. Dissecting mechanisms of miRNA regulation, however, is complicated by the generally poor annotation of miRNA primary transcripts (pri-miRNAs). Although some miRNAs are processed from well-defined protein coding genes, the majority of pri-miRNAs are poorly characterized noncoding RNAs, with incomplete annotation of promoters, splice sites, and polyadenylation signals. Due to the efficiency of DROSHA processing, the abundance of pri-miRNAs is very low at steady state, thereby complicating the elucidation of pri-miRNA structures. Here we describe a strategy to enrich intact pri-miRNAs and improve their coverage in RNA sequencing (RNA-seq) experiments. In addition, we outline a computational approach for reconstruction of pri-miRNA structures. This pipeline begins with raw RNA-seq reads and concludes with publication-ready visualization of pri-miRNA annotations. Together, these approaches allow the user to define and explore miRNA gene structures in a cell-type or organism of interest.

Key words microRNA, Primary transcript, pri-miRNA, RNA-seq, Transcriptome assembly

1 Introduction

Small noncoding RNAs function to regulate gene expression at transcriptional and post-transcriptional levels [1]. Among this class of regulatory RNAs, microRNAs (miRNAs) have emerged as key players in development and disease [2]. miRNAs act as negative regulators of gene expression, triggering the accelerated turnover and decreased translation of messenger RNAs (mRNAs) to which they bind [3]. The molecular mechanisms through which miRNAs regulate their targets have been extensively studied [4]. On the other hand, the lack of comprehensive annotation of primary miRNA gene structures has, in many cases, limited our understanding of how miRNA expression is regulated [5, 6].

Recent advances in high-throughput RNA sequencing (RNA-seq) have greatly improved transcriptome annotations [7, 8].

Ulf Andersson Vang Ørom (ed.), *miRNA Biogenesis: Methods and Protocols*, Methods in Molecular Biology, vol. 1823, https://doi.org/10.1007/978-1-4939-8624-8_1, © Springer Science+Business Media, LLC, part of Springer Nature 2018

Nevertheless, standard RNA-seq data are inadequate for reconstruction of miRNA gene structures. During miRNA biogenesis, miRNAs are initially transcribed as long primary transcripts (pri-miRNAs) that are rapidly processed in the nucleus by the microprocessor complex (consisting of DROSHA and DGCR8) [9]. Liberated precursor miRNA (pre-miRNA) hairpins are then exported to the cytoplasm where they are further processed by DICER prior to their loading into ARGONAUTE proteins, which they guide to target mRNAs to carry out their regulatory functions. As a result of the efficiency of this processing pathway, the pri-miRNA intermediates do not accumulate to levels that allow accurate and sensitive detection when using standard RNA-seq methodologies.

Here we provide a straightforward and highly effective strategy for genome-wide mapping of pri-miRNA structures, comprising both experimental and computational methods. First, a dominant-negative form of DROSHA (TN-DROSHA) [10] is introduced into cells to stall primary miRNA processing. This stabilizes the unprocessed pri-miRNAs, dramatically improving their coverage in RNA-seq libraries. Next, pri-miRNA structures are reconstructed using a streamlined workflow that incorporates StringTie [11], a recently developed algorithm that produces accurate pri-miRNA transcript assemblies [12].

The approach described here allows the genome-wide annotation of pri-miRNA transcripts in a cell-type of interest without any prior knowledge of miRNA gene structures, representing a major advantage over alternative methods such as rapid amplification of cDNA ends (RACE) and reverse-transcriptase polymerase chain reaction (RT-PCR). Furthermore, any cell line or organism in which TN-DROSHA can be expressed is eligible as the source of RNA. The method requires a relatively small amount of input nuclear RNA (0.1–4 μg) and the bioinformatic analysis of the RNA-seq results utilizes freely accessible computational software packages and assumes only basic UNIX skills. We therefore anticipate that application of this method will greatly improve our understanding of pri-miRNA structures and facilitate the investigation of miRNA regulation in diverse biological contexts.

2 Materials

Prepare all solutions using molecular biology grade reagents and ultrapure water with a sensitivity of 18.2 MΩ-cm at 25 °C. Prepare and store all reagents at room temperature (unless indicated otherwise). As a representative example, methods for performing this procedure in HEK293T cells are described here. Cell culture and transfection methodologies may need to be modified for use of other cell lines.

2.1 Reagents and Solutions

1. HEK293T cells or other cell line of interest.
2. Medium for cell culture: DMEM supplemented with 10% fetal bovine serum and antibiotics for HEK293T cells.
3. DNA plasmid pcDNA3.1/V5-His-TN-DROSHA: for expression of TN-DROSHA protein containing a C-terminally fused V5 and His tag [12].
4. DNA plasmid pcDNA5/FLAG-HA-DGCR8: for expression of DGCR8 protein containing an N-terminally fused FLAG and HA tag [12].
5. Transfection Reagent: Preferred transfection reagent for HEK293T cells.
6. Trypsin-EDTA.
7. Phosphate-buffered saline (PBS).
8. 10× Protease inhibitor stock solution: Dissolve EDTA-free Protease Inhibitor in water. Store the solution at −20 °C and use it within 12 weeks.
9. Buffer C: 10 mM Tris-HCl pH 7.5, 10 mM NaCl, 0.2 mM EDTA. Before use, add protease inhibitor from a 10× stock solution to a final concentration of 1×.
10. 10% (v/v) NP-40.
11. 0.1 M Dithiothreitol (DTT).
12. Buffer N: 20 mM Tris–HCl pH 7.5, 150 mM NaCl, 2 mM $MgCl_2$, 2 mM $CaCl_2$, 0.5% (v/v) NP-40, 0.1% (w/v) SDS. Before use, add DTT from a 0.1 M stock solution to a final concentration of 0.5 mM, and add protease inhibitor from a 10× stock solution to a final concentration of 1×.
13. V5 Tag Mouse Monoclonal Antibody (Invitrogen).
14. DGCR8 C-Terminal Antibody, Rabbit Polyclonal (Proteintech).
15. QIAzol Lysis Reagent (included in the Qiagen miRNeasy Mini Kit).
16. Qiagen miRNeasy Mini Kit.
17. RNase-free DNase.
18. Qubit RNA BR Assay Kit.
19. Illumina TruSeq RNA Library Preparation Kit v2.

2.2 Consumables and Equipment

1. Sterile, nuclease-free microcentrifuge tubes.
2. Sterile, nuclease-free pipet tips.
3. Sterile cell culture dishes.
4. Conical plastic centrifuge tubes.
5. Pipettes.

6. Vortex.
7. Refrigerated microcentrifuge and centrifuge.
8. Sonicator or Bioruptor.
9. Rocker shaker.
10. Thermal cycler.
11. Qubit Fluorometer.
12. Agilent 2100 Bioanalyzer.
13. Illumina HiSeq 2000 or other high-throughput sequencing system.
14. Hardware for bioinformatic analysis (64-bit computer running either Linux or Mac OS X; 4 GB of RAM (16 GB preferred)).

2.3 Software

1. fqtrim, for trimming/filtering of next-gen reads (http://ccb.jhu.edu/software/fqtrim/index.shtml).
2. TopHat2, for mapping spliced reads (http://ccb.jhu.edu/software/tophat/index.shtml).
3. StringTie, for assembling transcripts (https://ccb.jhu.edu/software/stringtie/).
4. gffread, for file conversion (http://ccb.jhu.edu/software/stringtie/gff.shtml).
5. UCSC Genome Browser (https://genome.ucsc.edu/).

3 Methods

3.1 Preparation of Nuclei and Evaluation of TN-DROSHA and DGCR8 Expression

1. Plate 4.5×10^6 HEK293T cells, or other cell line of interest, 24 h before transfection in a 10 cm tissue culture plate. On the day of transfection, mix 11.3 μg TN-DROSHA and 8.7 μg DGCR8 plasmids (molar ratio = 1:1) with 60 μL FuGENE HD Transfection Reagent according to the manufacturer's instructions. Incubate for 10–15 min, then add the FuGENE HD Transfection Reagent/DNA mixture to the cells to be transfected (*see* **Note 1**). Alternative transfection methods may be used if desired.
2. 48 h after transfection, gently wash the cells with PBS to remove dead cells and debris (*see* **Note 2**).
3. Harvest cells by trypsinization. Pellet the cells by spinning at $300 \times g$ for 5 min at room temperature. Wash the cell pellet once in ice cold PBS and keep the pellet on ice.
4. Tap the cell pellet loose, and gently resuspend the cells in 200 μL Buffer C per ~1×10^7 cells. Incubate the cells in Buffer C on ice for 10 min, allowing cells to swell (*see* **Note 3**).

5. To the swelled cells, add 10% NP-40 to a final concentration of 0.05% (e.g., 1 μL 10% NP-40 per 200 μL total volume). Incubate the cells on ice for 5 min or until the cells are lysed (*see* **Note 4**).
6. Centrifuge the lysed cells at 2500 × *g* for 10 min at 4 °C. Transfer the supernatant to a fresh tube and store on ice. This fraction is the cytoplasmic lysate.
7. Wash the pellet once in 100 μL Buffer C. Centrifuge again at 2500 × *g* for 5 min at 4 °C. Discard the supernatant and keep the pellet that contains nuclei.
8. To isolate total nuclear RNA for RNA-seq, continue to Subheading 3.2. To evaluate the expression of TN-DROSHA and DGCR8 in the cytoplasm and nucleus, cytoplasmic lysate from **step 6** is ready to be analyzed. For preparation of nuclear lysate, Buffer N is added to the nuclei from **step** 7 (100 μL Buffer N per ~1 × 10^7 nuclei). After sonication (*see* **Note 5**), the homogenized nuclear lysates can be analyzed for TN-DROSHA expression with anti-V5 antibodies and DGCR8 expression with anti-DGCR8 antibodies using standard immunoblotting methods (Fig. 1).

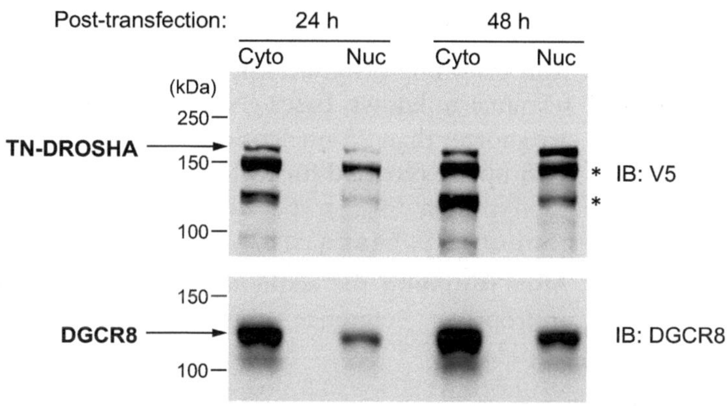

Fig. 1 Evaluation of TN-DROSHA/DGCR8 expression. Immunoblotting is used to confirm expression of TN-DROSHA and DGCR8 in transfected cells. As shown here for HEK293T cells, 48 h is often needed to accumulate high levels of nuclear TN-DROSHA/DGCR8 following transfection. Immunoblot analysis was performed by loading 25 μg of cytoplasmic (Cyto) and nuclear (Nuc) lysates on a NuPAGE 4–12% Bis-Tris protein gel (Invitrogen) with Precision Plus Protein Standards (Bio-Rad). Proteins were transferred to a nitrocellulose membrane and blocked with 5% non-fat milk/TBST for 1 h at room temperature. V5-His-TN-DROSHA was detected at ~160 kDa using anti-V5 tag antibody at a 1:4000 dilution and Flag-HA-DGCR8 was detected at ~120 kDa using anti-DGCR8 antibody at a 1:1200 dilution in 5% non-fat milk/TBST at 4 °C overnight on a rocker. Goat Anti-Mouse IgG-HRP or Goat Anti-Rabbit IgG-HRP secondary antibody (Bio-Rad) at 1:6000 dilution was used and chemiluminescent detection was performed using SuperSignal West Dura Extended Duration Substrate (Pierce). Asterisks denote DROSHA degradation products

3.2 RNA-Seq Library Preparation and Sequencing

1. Lyse the nuclei (from Subheading 3.1, **step** 7) in 700 μL QIAzol Lysis Reagent (up to 1×10^7 nuclei) by repeated pipetting until the nuclei are fully homogenized. Nuclear lysates can be stored at −80 °C for several months until ready for further processing.
2. Isolate total RNA using miRNeasy Mini Kit according to the manufacturer's instructions, including the on-column DNase digestion.
3. Measure RNA concentration with the Qubit RNA BR Assay Kit on a Qubit Fluorometer, and assess total RNA integrity using an Agilent 2100 Bioanalyzer (*see* **Note 6**). RNA can be stored at −80 °C until further use for at least a year.
4. Generate RNA-seq libraries using the TruSeq RNA Library Preparation Kit according to the manufacturer's instructions, and proceed to next-generation sequencing on a HiSeq 2000 or equivalent sequencing system using the 100 bp paired-end protocol (*see* **Note** 7). Longer paired-end reads may also be used.

3.3 Transcript Assembly

An overview of the RNA-seq data processing steps is shown in Fig. 2. These steps are performed using open-source software in a UNIX environment. While extensive bioinformatics expertise is not required, basic use of the UNIX command-line is assumed. The commands given here, prefixed with a "$" character, are meant to be executed from the UNIX shell.

1. Run trimming program fqtrim on raw FASTQ files to trim off terminal unknown bases (Ns) and to filter out the reads that are shorter than 25 nucleotides in length. Only reads which are ≥25 bp are retained for further analysis.
```
$ fqtrim -A -p 5 -l 25 -o trimmed.fq.gz
input_R1.fastq.gz,input_R2.fastq.gz
```
2. After trimming the sequencing reads, align the reads to the appropriate reference transcriptome and genome using TopHat2.
```
$ tophat2 -p 10 -o tophat -G known_genes.
gff3 --transcriptome-index=./tindex --li-
brary-type fr-firststrand hg19 input_
R1.trimmed.fq.gz input_R2.trimmed.fq.gz >&
run.tophat
```
3. Run StringTie to assemble RNA-Seq alignments into potential transcripts. The output file "accepted_hits.bam" from TopHat2 in **step 2** serves as the input to StringTie.
```
$ stringtie accepted_hits.bam -p 10 -S -g
0 -f 0.1 -o stringtie.gtf
```

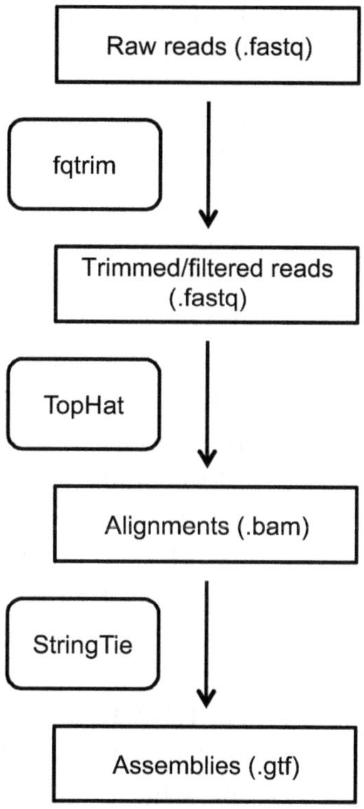

Fig. 2 An overview of the computational pipeline. Reads obtained by RNA-seq are first processed with fqtrim to trim off terminal unknown bases (Ns) and remove reads shorter than 25 nucleotides. The filtered reads are then mapped to the reference genome with TopHat. Mapped reads are provided as the input for StringTie, which assembles the transcripts that can be adapted for visualization on the UCSC genome browser

4. The output assembly is supplied as a GTF file. To display the assembly in the UCSC genome browser, run gffread to convert the file format.

```
$ gffread -T stringtie.gtf -o stringtie.
display.gtf
```

5. The converted GTF file can now be uploaded in the "Add Custom Tracks" page (accessible through the "add custom tracks" button; or http://genome.ucsc.edu/cgi-bin/hgCustom) in the UCSC Genome Browser, and added as a custom annotation track for visualization.

4 Notes

1. The procedures described here are optimized for expression of TN-DROSHA and DGCR8 in HEK293T cells. For successful expression of TN-DROSHA, the molar ratio of the plasmid pcDNA3.1/V5-His-TN-DROSHA to pcDNA5/FLAG-HA-

DGCR8, as well as the delivery methods (calcium phosphate, liposome transfection, or electroporation), must be optimized according to the cell type. Optimized transfection conditions for a limited set of human and mouse cell lines can be found in Chang et al. [12].

2. Dead or dying cells contain degraded RNA. It is essential to minimize the number of dead cells included in the RNA preparation steps since poor quality RNA will compromise sequencing data quality.
3. Cell swelling can be monitored by checking an aliquot of cells under light microscope.
4. It is important to ensure that the cells are properly lysed before proceeding to the next step. Trypan Blue solution can be added to an aliquot of cell suspension to check the lysis under light microscope.
5. For HEK293T cells, sonication is carried out using a Bioruptor in high power mode for 10 cycles (30 s ON, 30 s OFF) in a 4 °C water cooler.
6. An RNA Integrity Number (RIN) value greater than or equal to 8 is recommended.
7. To achieve high sensitivity and quality of pri-miRNA assemblies, at least 150 million paired-end reads per sample are recommended.

Acknowledgments

We thank Steven Salzberg and Mihaela Pertea at Johns Hopkins University for assistance with design of the bioinformatic pipeline and Stephen Johnson at UT Southwestern for software implementation and computational support. This work was supported by grants from the Cancer Prevention and Research Institute of Texas (RP160249) and the National Institutes of Health (R35CA197311). J. T. M. is an Investigator of the Howard Hughes Medical Institute.

References

1. Farazi TA, Juranek SA, Tuschl T (2008) The growing catalog of small RNAs and their association with distinct Argonaute/Piwi family members. Development 135(7):1201–1214. https://doi.org/10.1242/dev.005629
2. Bartel DP (2009) MicroRNAs: target recognition and regulatory functions. Cell 136(2):215–233. https://doi.org/10.1016/j.cell.2009.01.002
3. Pasquinelli AE (2012) MicroRNAs and their targets: recognition, regulation and an emerging reciprocal relationship. Nat Rev Genet 13(4):271–282. https://doi.org/10.1038/nrg3162
4. Gurtan AM, Sharp PA (2013) The role of miRNAs in regulating gene expression networks. J Mol Biol 425(19):3582–3600. https://doi.org/10.1016/j.jmb.2013.03.007

5. Schanen BC, Li X (2011) Transcriptional regulation of mammalian miRNA genes. Genomics 97(1):1–6. https://doi.org/10.1016/j.ygeno.2010.10.005
6. Winter J, Jung S, Keller S, Gregory RI, Diederichs S (2009) Many roads to maturity: microRNA biogenesis pathways and their regulation. Nat Cell Biol 11(3):228–234. https://doi.org/10.1038/ncb0309-228
7. Martin JA, Wang Z (2011) Next-generation transcriptome assembly. Nat Rev Genet 12(10):671–682. https://doi.org/10.1038/nrg3068
8. McGettigan PA (2013) Transcriptomics in the RNA-seq era. Curr Opin Chem Biol 17(1):4–11. https://doi.org/10.1016/j.cbpa.2012.12.008
9. Ha M, Kim VN (2014) Regulation of microRNA biogenesis. Nat Rev Mol Cell Biol 15(8):509–524. https://doi.org/10.1038/nrm3838
10. Heo I, Joo C, Cho J, Ha M, Han J, Kim VN (2008) Lin28 mediates the terminal uridylation of let-7 precursor MicroRNA. Mol Cell 32(2):276–284. https://doi.org/10.1016/j.molcel.2008.09.014
11. Pertea M, Pertea GM, Antonescu CM, Chang TC, Mendell JT, Salzberg SL (2015) StringTie enables improved reconstruction of a transcriptome from RNA-seq reads. Nat Biotechnol 33(3):290–295. https://doi.org/10.1038/nbt.3122
12. Chang TC, Pertea M, Lee S, Salzberg SL, Mendell JT (2015) Genome-wide annotation of microRNA primary transcript structures reveals novel regulatory mechanisms. Genome Res 25(9):1401–1409. https://doi.org/10.1101/gr.193607.115

Chapter 2

Identifying Pri-miRNA Transcription Start Sites

Georgios Georgakilas, Nikos Perdikopanis, and Artemis G. Hatzigeorgiou

Abstract

MicroRNAs (miRNAs) are small non-coding RNAs that can regulate gene expression playing vital role in nearly all biological pathways. Even though miRNAs have been intensely studied for more than two decades, information regarding miRNA transcription regulation remains limited. The rapid cleavage of primary miRNA transcripts (pri-miRNAs) by Drosha in the nucleus hinders their identification with conventional RNA-seq approaches. Identifying the transcription start site (TSS) of miRNAs will enable genome-wide identification of their expression regulators, including transcription factors (TFs), other non-coding RNAs (ncRNAs) and epigenetic modifiers, providing significant breakthroughs in understanding the mechanisms underlying miRNA expression in development and disease. Here we present a protocol that utilizes microTSS, a versatile computational framework for accurate and single-nucleotide resolution miRNA TSS predictions as well as miRGen, a database of miRNA gene TSSs coupled with genome-wide maps of TF binding sites.

Key words microRNA, miRNA, TSS, Transcription start site, TF, Transcription factor, pri-miRNA, RNA-seq, DNase, ChIP-seq, Histone marks, Machine learning

1 Introduction

1.1 MicroRNA Biogenesis and Function

MicroRNAs (miRNAs) are short (~22 nts) non-coding RNA (ncRNA) molecules that regulate gene expression of protein-coding genes through transcript degradation and/or translation suppression. Even though miRNAs have been intensely studied since 1993 [1], many aspects of miRNA transcription regulation remain to be discovered. The transcription of most mammalian miRNAs is driven by RNA Polymerase II (Pol2) resulting in the formation of capped, polyadenylated and in many cases spliced transcripts, named primary-microRNAs (pri-miRNAs). Over the years, it has become apparent that miRNA genes share the same mechanisms of transcription and post-transcriptional processing with protein-coding genes. More than half of mammalian miRNAs are encoded near other miRNA loci. These clustered miRNAs are derived from a single polycistronic transcription unit [2, 3]. In general, miRNAs can be divided into two categories depending on

Ulf Andersson Vang Ørom (ed.), *miRNA Biogenesis: Methods and Protocols*, Methods in Molecular Biology, vol. 1823, https://doi.org/10.1007/978-1-4939-8624-8_2, © Springer Science+Business Media, LLC, part of Springer Nature 2018

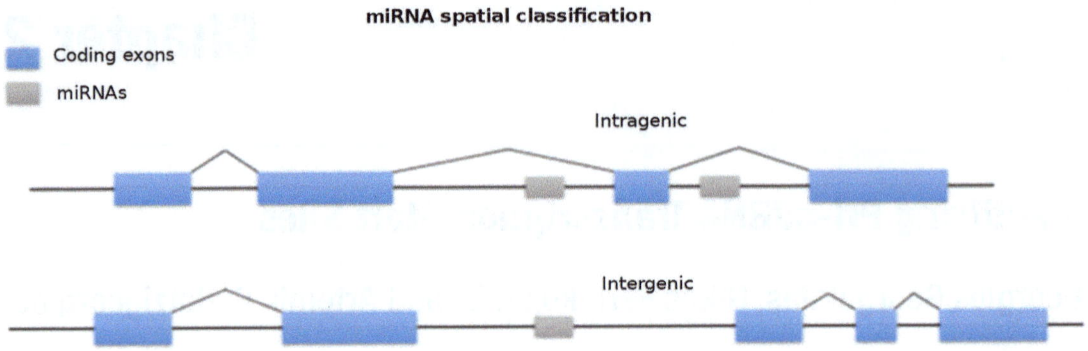

Fig. 1 Spatial classification of miRNAs relative to protein-coding genes. This figure has been derived from [4]

their location relative to protein-coding genes. Intragenic miRNAs are in either the intronic or the exonic part of protein-coding genes. On the other hand, intergenic miRNAs are encoded in individual miRNA genes located in the genomic space between coding loci (Fig. 1).

Most intragenic miRNAs share the same promoter with the host gene; however, in some cases they have their own regulatory loci residing in upstream intronic regions. Transcription is an integral part in the mechanism of miRNA biogenesis regulation. A plethora of Pol2-associated transcription factors are involved in the regulation of miRNA genes. The initial products of Pol2-mediated miRNA gene transcription are pri-miRNAs whose length varies between a few hundred to several hundred kilobases. Pri-miRNAs contain hairpin-like structures (Fig. 2), named precursor-microRNAs (pre-miRNAs), which can be identified by the nuclear RNase III type protein Drosha [5].

This step of miRNA biogenesis pathway is localized in the nucleus and requires another protein, named DGCR8, in order for a large dimer known as Microprocessor complex to be formed [6]. Typically, pre-miRNAs include the mature miRNA and its complementary sequence, the stem which is approximately 33 nts in length and flanking single stranded RNAs (ssRNAs). Drosha can identify and cleave the substrate ~11 bp away from the ssRNA–stem junction. This process is catalyzed by the interaction between DGCR8, the stem loop, and the ssRNA segments [7]. There is increasing evidence that pri-miRNA processing may be a co-transcriptional process. In cases where the hairpin structure is located inside exonic region of protein-coding genes, the cleavage by the Microprocessor complex can induce reduced protein production. Pre-miRNAs, released from primary transcripts in the nucleus, are subsequently exported to the cytoplasm by a member of the nuclear transport receptor family of proteins, named EXP5. Upon reaching cytoplasm, RNase III type protein Dicer cleaves ~22 nts away from the stem base of the hairpin structure releasing

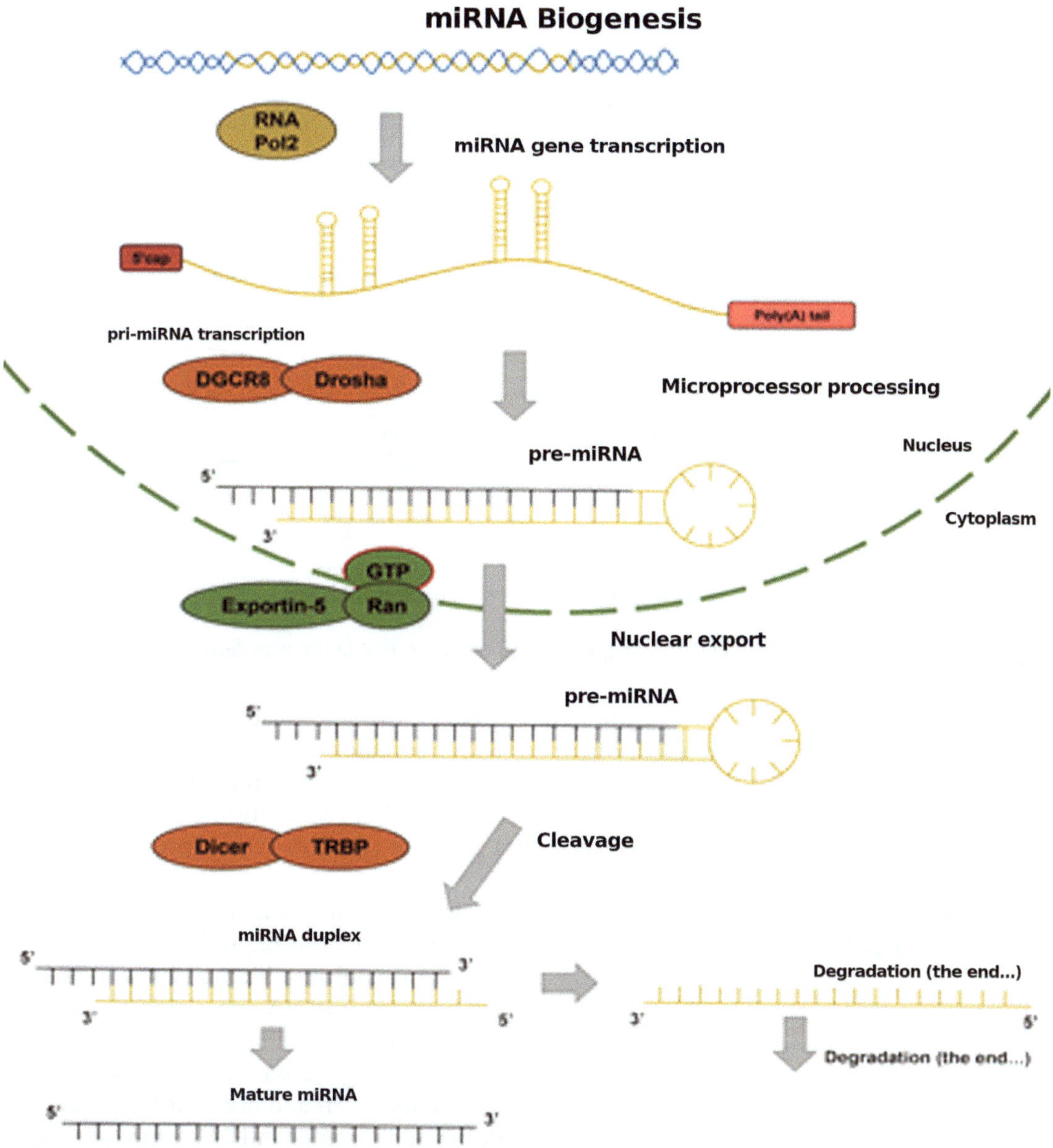

Fig. 2 The miRNA biogenesis pathway. This figure was derived from [4]

miRNA duplexes of approximately 18–23 nts in length [8]. One strand of the produced RNA duplex will be subsequently loaded to the RNA-induced silencing complex (RISC) while the other strand is typically degraded. In some cases, some pre-miRNAs produce mature sequences from both strands that survive and are functional in comparable frequencies [9].

In the final step of miRNA biogenesis pathway, the mature products are loaded onto complexes, either referred to as ribonucleic proteins (RNPs) or RISCs, following Dicer processing and

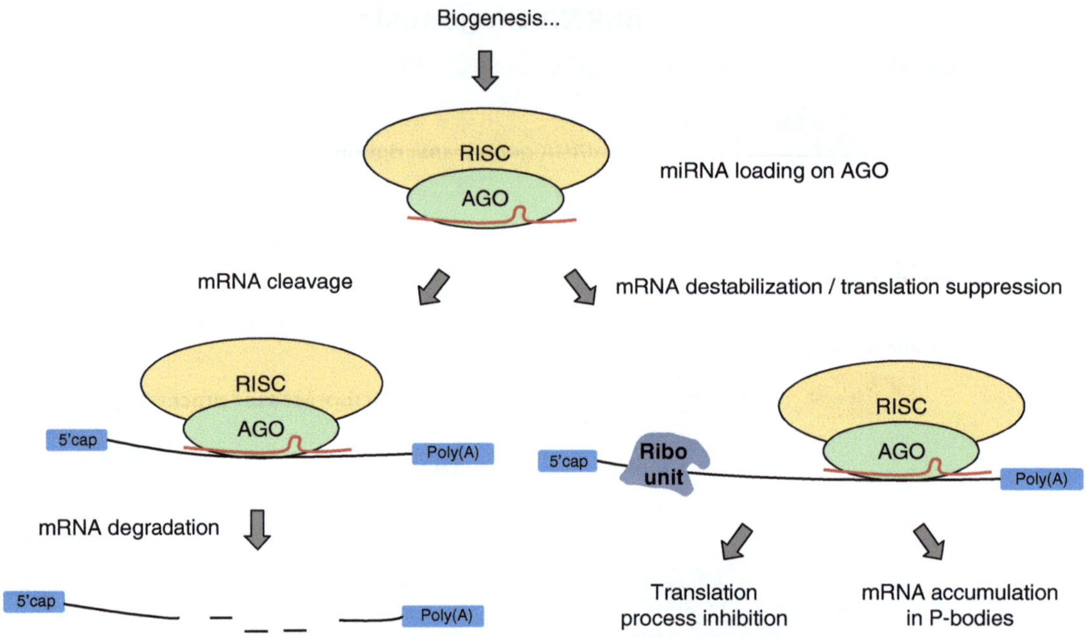

Fig. 3 Gene expression regulation by miRNA-guided RISC complex. This figure was derived from [4]

removal of the stem loop from the hairpin structures (Fig. 3). The major component of RISC complex is the Argonaute family of proteins [10] which utilizes the loaded mature miRNA sequence and its complementarity to miRNA recognition elements (MREs) located on mRNAs as a guide to regulate gene expression.

1.2 Implications of miRNAs in Disease

Since their first discovery in 1993, miRNAs have been vigorously researched for their implication in various physiological and pathological states in a plethora of organisms. Numerous studies have reported a tight interplay between miRNA expression and important mechanisms responsible for the development of various species. Small alterations in miRNA biogenesis pathway, in example the inhibition of critical components of pri- and pre-miRNA processing machineries, result in global inhibition of miRNA expression leading to lethality in early embryonic stages [11]. The role of miRNAs in differentiation is not limited to embryonic development. In example, conditional Dicer knockout mice models in myogenic tissues result in abnormal morphology of muscle fibers [12]. The normal development of the hematopoietic lineage has been found to depend on proper expression of certain miRNAs regulated by cell type specific transcription factors (TFs). GATA1 activates the transcription of miR-451 and miR-23 which are responsible for the final differentiation of erythroid progenitors [13].

Since the early stages of miRNA research, cancer has been the most prominent among human diseases with a clear role for miRNA regulation. Initial evidence involved the observation of lower miR-15 and miR-16 abundance in 65% of B-cell chronic lymphoma leukemia patients [14]. Subsequent expression profiling studies further established the connection between aberrant miRNA expression patterns of miR-125b, miR-145, miR-21, and miR-155, and the increased risk of breast cancer [4]. Upregulation of miR-155 and downregulation of let-7a have been correlated with low survival rates of lung cancer patients [15]. Additional studies highlight the imbalance between cell death and proliferation during the development of various types of cancer such as hepatocellular carcinoma [16], breast cancer [17], and adenocarcinoma [18].

The type of pathological conditions associated to miRNAs is not limited to cancers. Many immune-related diseases such as fatty liver disease, systemic lupus erythematosus, type I/II diabetes, and multiple sclerosis have shown established connections with specific miRNAs. Numerous mature miRNAs were identified as signatures by analyzing the expression profiles of healthy controls and relapsing multiple sclerosis patients [19]. In two independent studies including hundreds of systemic lupus erythematosus patients and healthy controls, the decreased miR-146a expression was strongly correlated with increased risk for the disease among European and Asian populations [20]. The pathogenesis of neuronal degeneration such as Parkinson and Alzheimer remains poorly understood. There is a progressively growing number of studies attempting to shed light on the implication of miRNAs in such diseases. In example, expression deregulation of miR-133b might contribute to the progression of Parkinson's disease, since the miR-133b-PIXT3 feedback loop is essential for maintaining dopaminergic neurons in the brain [21]. The first cases of viral-encoded miRNAs have been derived from a Burkitt's lymphoma cell line which was infected by Epstein–Barr virus [22]. Since then, bioinformatics and cloning approaches we utilized to identify viral miRNAs in polyoma virus [23], adenovirus [24], and several subtypes of the herpes viruses [25]. The function of viral miRNAs mainly focuses on targeting host genes that assist the cell to enter the apoptotic cycle.

Even though significant progress has been achieved for the identification of miRNA function, information regarding miRNA transcription regulation remains significantly limited. Such knowledge will enable the genome-wide identification of miRNA expression regulators, including transcription factors (TFs), other non-coding RNAs and epigenetic modifiers; providing significant breakthroughs in understanding the mechanisms underlying miRNA expression in development and disease.

1.3 The Field of microRNA Transcription Start Site Prediction

Initial attempts to characterize the miRNA promoterome include methods that utilize DNA sequence features (e.g., overrepresented k-mer distributions and CpG islands) [26–29]. However, such approaches suffer from increased false positive rates. In addition, their predictions are outside of any cell-specificity context which in addition to the number of false positives renders their validation not feasible and their usage impractical.

The second generation of miRNA TSS prediction algorithms took advantage of the first wave of high-throughput Next Generation Sequencing (NGS) experimental assays (i.e., chromatin immunoprecipitation sequencing, ChIP-seq) that provided quantitative measures of protein::DNA interactions. This family of algorithms was based on the integration of Polymerase II and histone H3K4me3 ChIP-seq signal, both marks of active/poised proximal regulatory elements [30–32]. The false positive rate was dramatically reduced, and cell type specific predictions were introduced for the first time. However, the resolution of the provided miRNA TSS predictions (often distributed over several kilobases [2]) remained a limiting factor due to the epigenetic complexity of multi-gene loci.

The latest generation of algorithms incorporate a broad spectrum of -omic datasets such as CAGE, RNA-, TSS- and/or ChIP-seq, and combine them with generic DNA features (overrepresented k-mer distributions, CpG island and sequence conservation). CAGE or TSS oriented algorithms [29, 33, 34] operate by merging high-throughput libraries from a wide array of cell lines and tissues which results in multiple predictions per miRNA. Additionally, the nature of 5′ cap identification protocols such as CAGE and TSS-seq skews the distribution of predictions towards non-TSS-related molecules and phenomena such as recapping events and byproducts of the splicing machinery.

MicroTSS [2] is the first algorithm to incorporate Pol2 and H3K4me3 ChIP-seq as well as RNA-seq and DNase-seq derived footprints and achieves the best of both worlds, a significant breakthrough in performance metrics (Table 1) and single-nucleotide resolution for cell-specific miRNA TSS prediction (Fig. 4).

Taking advantage of the recent improvements in the NGS field, microTSS utilizes deeply sequenced (>100 million reads) RNA libraries and detects putative 5′ ends of pri-miRNA transcripts (Fig. 2). In subsequent steps, these putative TSSs are filtered based on Support Vector Machines (SVM) models trained on the signal distribution of Pol2 and H3K4me3 ChIP-seq data as well as TF footprints as identified by DNase-seq around protein-coding TSSs. The performance of microTSS and other in silico miRNA TSS detection methods was validated against TSSs identified using mESC RNA-seq data derived from a Drosha-null/conditional-null mouse model generated with the COIN methodology [2, 35]. Fourty-four out of fourty-seven miRNA TSSs characterized in Drosha depleted ESCs were correctly identified by microTSS (Table 1). It is evident that the multi-tissue data incorporation feature of other algorithms greatly limits their performance.

Table 1
Comparison between microTSS, S-Peaker, Marson et al., and PROmiRNA in terms of precision and sensitivity using Drosha-null-validated miRNA TSSs

Algorithms precision and sensitivity at 1 kbp distance threshold from validated TSSs in mESC		
	mESCs (*N* = 47)	
	Sensitivity	**Precision**
Marson et al.	54% (20/37)	64.5% (20/31)
PROmiRNA	78.7% (37/47)	25.4% (95/373)
S-Peaker	76.5% (36/47)	18.8% (77/409)
microTSS	93.6% (44/47)	100% (44/44)

Only predictions in distance <1000 bp are considered true. Marson et al. supports 31 out of the 47 experimentally validated TSSs and provides 33 predictions, microTSS supports 44 TSSs by providing one prediction per miRNA while both PROmiRNA and S-Peaker support 47 TSSs and provide 373 and 409 predictions, respectively. This table has been adapted from [15]

1.4 Resources of miRNA Expression Regulators

A common characteristic between available miRNA transcription regulation repositories is the lack of supporting accurate miRNA TSSs. The field of miRNA research and especially the effort of incorporating miRNAs into the context of gene regulatory networks are highly dependent on the accurate miRNA promoterome characterization. The majority of implementations that try to infer TF::miRNA interactions cluster pre-miRNAs into transcriptional units by utilizing heuristic methods based on inter-miRNA distances. However, pri-miRNAs can span dozens of kilobases and such approaches can often be misleading. Other databases utilize the predictions of early miRNA TSS identification algorithms, most of which support low resolution predictions derived from wide transcription signals, such as histone marks.

MiRGen [36] was initially released in 2007 and focused on exploring the association between the genomic context of miRNAs and their function, providing spatial annotation of miRNAs as well as validated and in silico-derived connections between miRNAs and their target genes. MiRGen v2 [37] attempted to expand the scope of the database by incorporating information related to pri-miRNA expression regulation. Promoter regions of miRNAs were derived from computational and experimental sources, while TF::miRNA associations were calculated by position weight matrix (PWM) mediated motif scanning of the promoter regions.

ChIPBase [38] and CircuitsDB [39] group pre-miRNAs into transcriptional units and consider the cluster's TSS as the 5′ end of

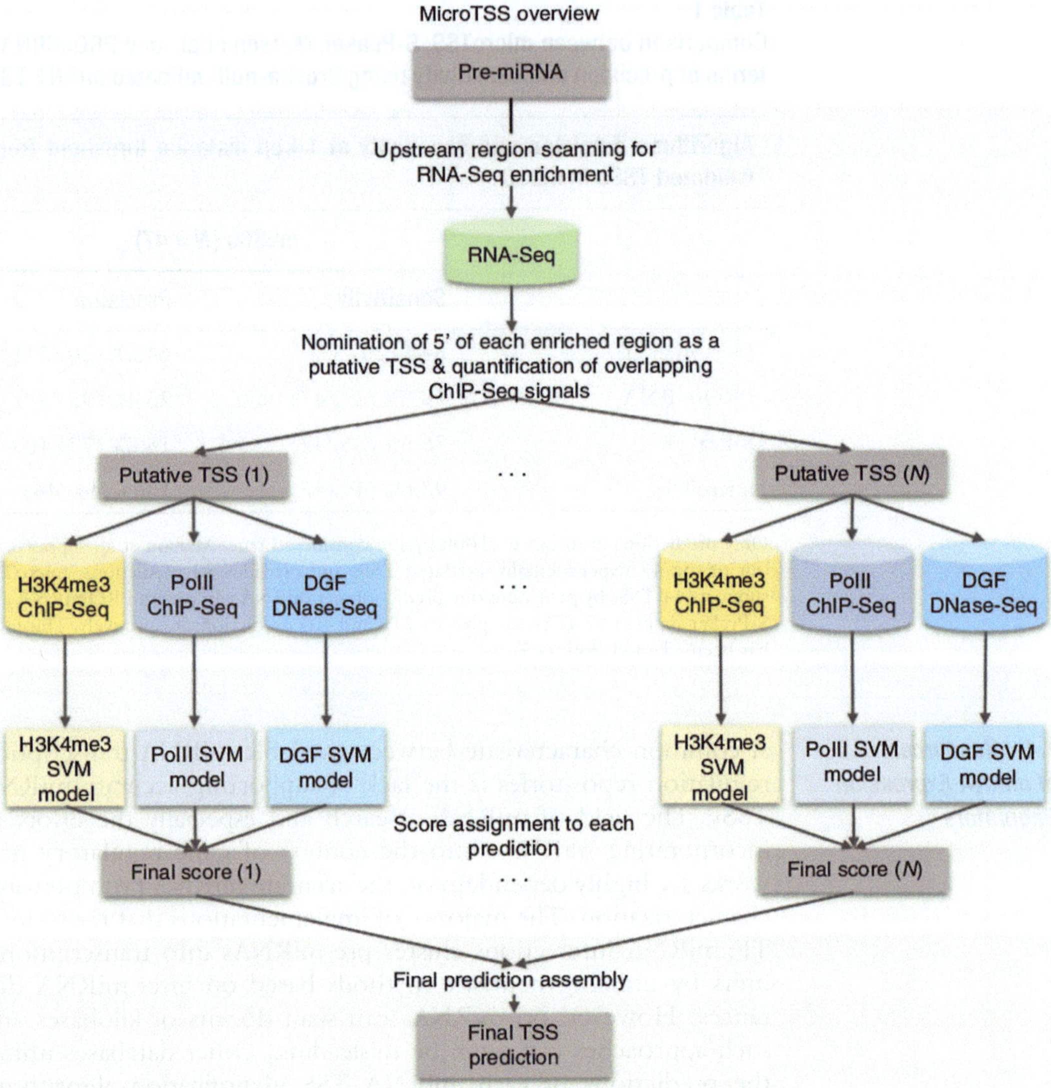

Fig. 4 MicroTSS workflow diagram. The first step in microTSS is to identify RNA-seq enriched regions. This is accomplished by utilizing a sliding window initialized at the pre-miRNA genomic location. The window scans the pre-miRNA upstream region based on a user defined distance threshold. The 5′ end of the identified RNA-seq enriched loci serves as a TSS candidate. Subsequently, the number of H3K4me3/Pol2 ChIP-seq reads and DNase footprints overlapping the area surrounding each candidate TSS constitute features of three distinct SVM models (one for each type of transcription mark). The output of each SVM model is summed up to produce the final score of each candidate TSS. This figure was derived from [15]

the first cluster member. ChIPBase integrates TF ChIP-seq libraries derived from ENCODE while CircuitsDB infers TF::miRNA interactions by motif scanning of a limited region surrounding the miRNA clusters.

TMREC [40] and TransmiR [41] are databases that utilize text mining on ~100 and 5000 publications, respectively, in order to extract TF–miRNA interactions in 21 diseases from 16 organisms. The quality of the hosted information is highly dependent on the curation process of the text mining results.

TSmiR [42] provides interaction maps between TFs, miRNAs and their target genes in multiple human tissues. Their TSSs have been obtained from already published in silico and experimental methodologies. ChIP-seq-derived TF binding sites were downloaded from the ENCODE project repository for each of the studied tissues.

Despite the progress in unveiling the mechanisms of miRNA transcription regulation, most of existing studies for the identification of miRNA TSSs rely on low accuracy experimental techniques, in silico algorithms that provide low resolution/high false positive rate predictions and heuristics. In addition, the assembly of TF–miRNA interactions frequently is based on literature text-mining, TF motif-assisted scanning of the promoter regions and ChIP-seq, which limits the search in one TF per experiment. Importantly, previously available genome-wide NGS-based miRNA TSS identification techniques have low accuracy and a low signal to noise ratio [2].

DIANA-miRGen v3.0 [43] attempted to establish for the first time accurate and extensive maps of miRNA promoters and to remove the obscurity that surrounds miRNA transcription regulation by providing an accurate genome-wide map of TF::miRNA interactions for a plethora of tissues and cell lines in human and mouse. MicroTSS [2] facilitated the identification of 276 tissue/cell-line specific TSSs for 428 miRNA precursors. More than 200 TF PFMs were combined with in-house assembled RNA-seq expression profiles and TF footprints derived from DNase-seq, to create sets of motifs specific to each of the studied tissues and cell lines. This wealth of information can be accessed through an intuitive and easy-to-use interface (Fig. 5) that incorporates rich meta-data regarding the function of miRNAs and TFs as well as their implication in physiological conditions and diseases while at the same time providing seamless interconnection between miRGen v3.0 and other DIANA resources.

2 Materials

DIANA-miRGen v3.0 and microTSS v1.0 are both part of the DIANA-suite (www.microrna.gr), a set of state-of-the-art tools and databases aiming to interpret and archive data in a systematic framework ranging from the analysis of gene expression regulation (DIANA-mirExTra v2.0 [44]), the annotation of miRNA targets (DIANA-microT-CDS [45], TarBase v8.0 [46], and LncBase [47]) to the interpretation of the role of ncRNAs in various diseases and pathways (DIANA-miRPath [48]) (*see* **Notes 1** and **2**).

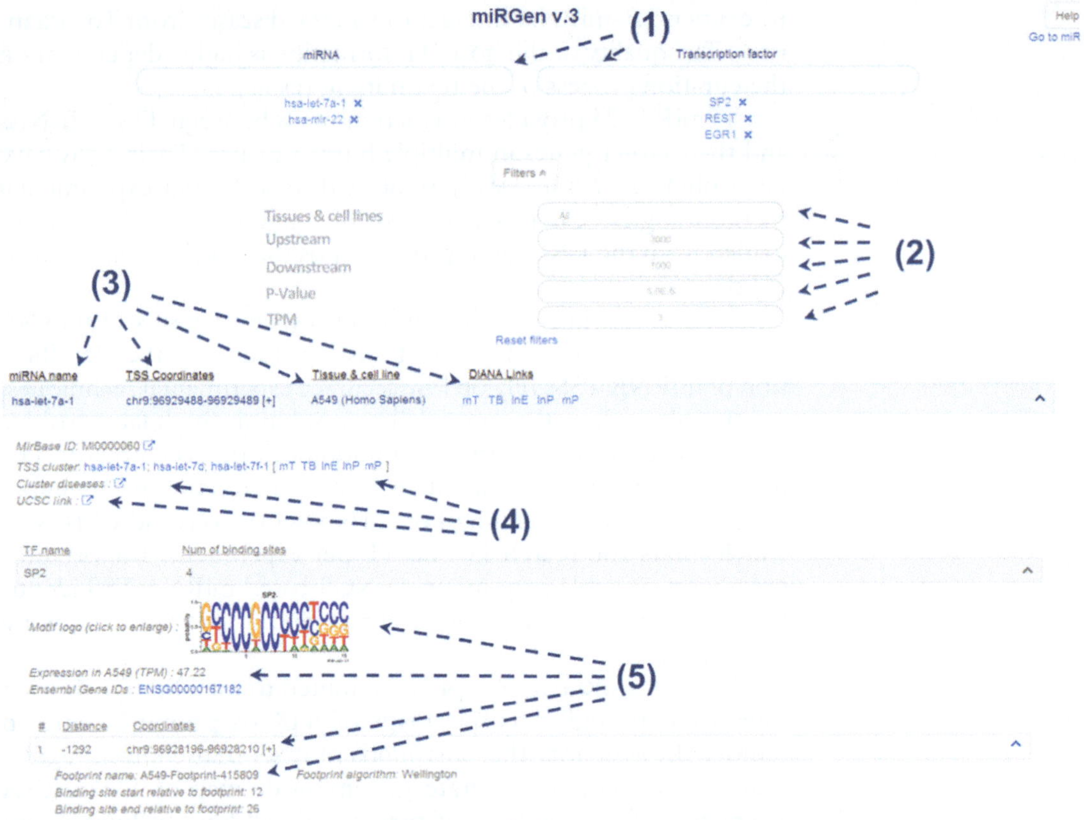

Fig. 5 Main miRGen interface panels: (1) Input fields for pre-miRNA and/or TF names and (2) filters pane for query customization. Query execution returns information about the genomic coordinates and tissue specificity of the predicted miRNA TSS (3), miRNA::disease associations (4), and TFs regulating its expression (5). Links for external databases are also provided. This figure has been derived from [16]

To access miRGen v3.0 and other components of the DIANA-suite, users only need access to a personal computer or mobile device with internet connection and web browser support (all main browsers are supported, e.g., IE, Firefox, Chrome, Safari, Opera).

microTSS v1.0 is an *offline* computational framework written in Perl and requires access to a personal computer or workstation with a minimum set of requirements in hardware and software (Table 2).

3 Methods

The presented protocol utilizes microTSS v1.0 framework as well as miRGen v3.0 repository to identify pri-miRNA transcription start sites TSS and uncover TF driving miRNA expression. MicroTSS and miRGen are completely autonomous and can be used independently.

Table 2
Minimum requirements in hardware and software required for running microTSS

Hardware	At least 8 gigabytes of RAM recommended. Multi-threading is also supported
Operating system:	Any Linux distribution would suffice. Extensively tested on Ubuntu 14.01 LTS
Prerequisite software:	• Perl version 5 and above with the following modules: 1. Getopt::Long (it usually comes with the default perl installation) 2. File::Basename (it usually comes with the default perl installation) 3. File::Spec (it usually comes with the default perl installation) 4. PerlIO::gzip 5. Set::IntervalTree 6. Parallel::ForkManager All modules can be downloaded by CPAN • SVM module. LIBSVM v3.17 is required and should be in the PATH. LIBSVM can be found in *http://www.csie.ntu.edu.tw/~cjlin/libsvm/*. • BEDTools suite is required and should be in the PATH. BEDTools can be found in *https://github.com/arq5x/bedtools2*
Data sources:	To run microTSS, user need to supply the following data • File containing RNA-seq read • File containing the H3K4me3 reads • File containing the polymerase II data • File containing the DGF transcription factor binding sites All files must be supplied on BED format compressed using gzip or uncompressed (bed.gz or bed). In case of data files in BAM format, it is highly recommended to use the BAM to BED utility provided by microTSS framework, located in directory *microTSS_v1.0/src/tools/*

3.1 Applying microTSS

3.1.1 Installation

After installing all prerequisite software (libraries and external tools) described in the Subheading 2, users need to navigate to *http://www.microrna.gr/microTSS* and download the source code of microTSS. Following extraction users will observe microTSS's directory structure (Fig. 6).

Details about the purpose of each subdirectory, required input files, and usage parameters can be found in the README file included in the source code archive.

3.1.2 Usage

MicroTSS is a command line application with multiple optional and required parameters. The software comes with a set of pre-defined parameters optimized to work well under most scenarios (Table 3).

microTSS provides files from intermediate steps of execution (Table 4) to assist users in fine tuning the analysis. The final list of miRNA TSS predictions is included in "*tss_predictions.bed*." Every step of the microTSS analysis provides bed-like formatted files that can be readily used in meta-analyses and loaded on any genome browser or web server.

```
├── data
│   ├── human
│   └── mouse
├── example
├── microTSS.pl
└── src
    ├── lib
    ├── model
    └── tools
```

Fig. 6 microTSS directory structure. MicroTSS.pl is the main script needed to run the algorithm. Sample data are provided in the example directory

3.1.3 Use Case

Users can evaluate microTSS using sample RNA-seq (GSM973235), H3K4me3 ChIP-seq (GSM723017), Pol2 ChIP-seq (GSM723019), and Dnase-seq derived TF footprints (GSE40869) data located in the directory "*microTSS_v1.0/example*." The following code snippet shows the parameters used to apply microTSS on the sample datasets.

```
perl microTSS.pl    --file_to_scan=example/let7d_cluster.bed
                    --result_directory_id=example_run
                    --svm_score_threshold=1.5
                    --max_processes=1
                    --strand_mode=stranded_RF
                    --species=mm10 --rnaseq_directory=example/
                    --histone_chipseq_file=example/H3K4me3.mESC.mm10.bed.gz
                    --polymerase_chipseq_file=example/Pol2.mESC.mm10.bed.gz
                    --dgf_dnase_file=example/DGF.mESC.mm10.bed.gz
```

After completion of the algorithm's execution, all output files can be found in the directory "microTSS_v1.0/data/mouse/mm10/example_run" (as defined by result_directory_id parameter). Upon loading some of the intermediate files as well as the final miRNA TSS prediction file, we can observe that microTSS supports a single TSS prediction for the let-7 miRNA cluster (Fig. 7).

3.2 Utilizing miRGen

DIANA-miRGen v3.0 is a repository for miRNA transcription regulation and comprises accurate cell-line specific miRNA TSSs and genome-wide maps of TFBS. MicroTSS has been applied on RNA-seq, ChIP-seq, and DNase-seq data derived from the ENCODE [49] consortium portal in order to extract accurate miRNA transcription start site predictions. In order to assemble a genome-wide map of transcription factor binding sites in the studied tissues and cell lines, approximately 4.1 billion DNase-seq reads [49, 50] were analyzed. HOMER [51] was utilized for the detection of genomic regions enriched in DNase-seq signal (hotspots) with a false discovery rate threshold of 1%. Hotspots were subsequently processed with Wellington algorithm [52] to identify transcription factor footprints using a p-value threshold of 10^{-8}.

Table 3
microTSS list of parameters

Optional parameters	Description	Default value
--file_to_scan	Name of the file containing the pre-miRNA annotation	Default annotation
--result_directory_id	Name of the result directory	A random id will be generated
--max_processes	Number of processes for parallel execution of the algorithm	Default = 1 process
--upstream_search_region	Upstream region to be scanned for RNA-seq enrichment	Default = 250,000 base pairs
--window_size	Size of the scanning window	Default = 30 base pairs
--window_step	Scanning window step in bp	Default = 5 base pairs
--gap_size	Maximum required distance between discovered RNA-seq enriched islands to be considered as a single island	Default = 200
--window_score_threshold	Minimum number of overlapping reads required to form an RNA-seq enriched island	Default = 5
--svm_score_threshold	SVM score threshold for filtering TSS predictions, 0–3	Default = 1.5
--in_between_prediction_cov	RNA-seq coverage threshold between TSS predictions for a given pre-miRNA	Default = 0.1.
--strand_mode	Strand mode allowing of the RNA-seq experiment. Choose between stranded_RF/stranded_FR/unstranded	Default = unstranded
Mandatory parameters	**Description**	
--species	Choose between mm 9, mm 10, and hg 19. For additional species, please contact us	
--rnaseq_directory	This should be the directory containing the RNA-seq reads in bed format. The algorithm expects to find one file per chromosome (compressed with gzip or not), i.e., chr13.bed.gz or chr13.bed. It is highly recommended to use the bam to bed utility provided by microTSS which is located in microTSS_v1.0/src/tools/	
--histone_chipseq_file	File containing the H3K4me3 reads in bed format. It can be either compressed with gzip or not, i.e., H3K4me3.bed.gz or H3K4me3.bed	
--polymerase_chipseq_file	File containing the polymerase II reads in bed format. The file can be either compressed with gzip or not, i.e., PolII.bed.gz or pol II.bed	
--dgf_dnase_file	File containing the DGF transcription factor binding sites in bed format. It can be either compressed with gzip or not, i.e., DGF.bed.gz or DGF.bed	

Certain parameters such as –window_step, —window_size, and –gap_size can greatly affect the results

Table 4
List of microTSS output files

Output file	Description
rnaseq_enriched_islands.bed	This is the output of the first step of the algorithm which is the rnaseq_enriched_region_caller.pl script. It contains all the identified RNA-seq enriched regions upstream of the regions included in --file_to_scan or the default miRNA annotation
rnaseq_enriched_islands_scored_*.txt	This is the output of the second step of the algorithm which is the score_rnaseq_enriched_regions.pl script. Each RNA-seq enriched region identified in the previous step is scored based on H3K4me3 and Pol2 ChIP-seq and DGF TF binding sites as identified by DNase-seq. * refers to histone, polymerase, and footprint
rnaseq_enriched_islands_libsvmScored_*.txt	Same as above but this is the output of each SVM model. * refers to histone, polymerase, and footprint
tss_predictions.bed	This is the output of the third step of the algorithm which is the promoter_caller.pl & process_final_predictions.pl. This contains the final TSS predictions of microTSS. Note that for clustered miRNAs there will be one prediction and the names will be delimited with ";"

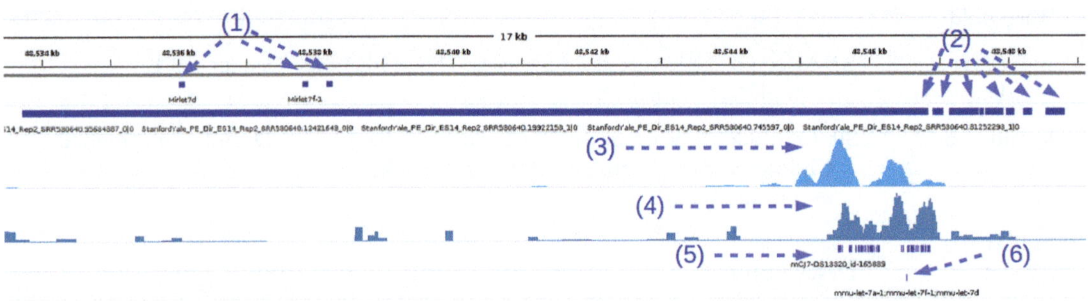

Fig. 7 Input data (RNA-seq, H3K4me3, Pol II, DNase-seq TF footprints) processed by microTSS coupled with the predicted TSS for the let-7 cluster. (1) miRNA cluster located in the antisense strand. (2) RNA-seq enriched regions; their 5′ end signifies putative TSSs for this cluster. (3) H3K4me3 ChIP-seq signal. (4) Pol2 ChIP-seq signal. (5) TF footprints as identified by DNase-seq. (6) Predicted TSS of the let-7 cluster by microTSS

3.2.1 miRGen Use Case

As a use case scenario, we will explore the function of miRNAs expressed in cardiac tissue with TFs specific for the same tissue [53, 54] (e.g., GATA4, SRF, and MEF2A). These regulatory proteins regulate the expression of multiple miRNAs, some of these are highly important in cardiac pathways. In example, let-7c-1/miR-99a cluster controls cardiomyogenesis [55], miR-1a-1 regulates cell proliferation and cardiac function [56], miR-133a enhances the protective capacity of cardiac progenitor cells after myocardial infarction [57], miR-23a/miR-24-2/miR-27a cluster is involved in cardiac hypertrophy [58] as well as heart fibrosis after myocardial infarction [59] and deregulation of miR-30 family members has been associated to various heart-related pathologies [60, 61].

To reproduce this study users have to navigate to miRGen v3.0 starting page (www.microrna.gr/mirgen) and fill GATA4, SRF, and MEF2A in *transcription factor* form field (Fig. 8).

Subsequently select Heart (Mus musculus) from *Tissues & cell lines* named filter on the left pane of the form. Users can define the length of the region surrounding each TSS (upstream and downstream) that will be investigated for TF binding sites as well as threshold values for binding sites strength (p-values calculated from FIMO) and minimum acceptable expression level for each TF (Fig. 9).

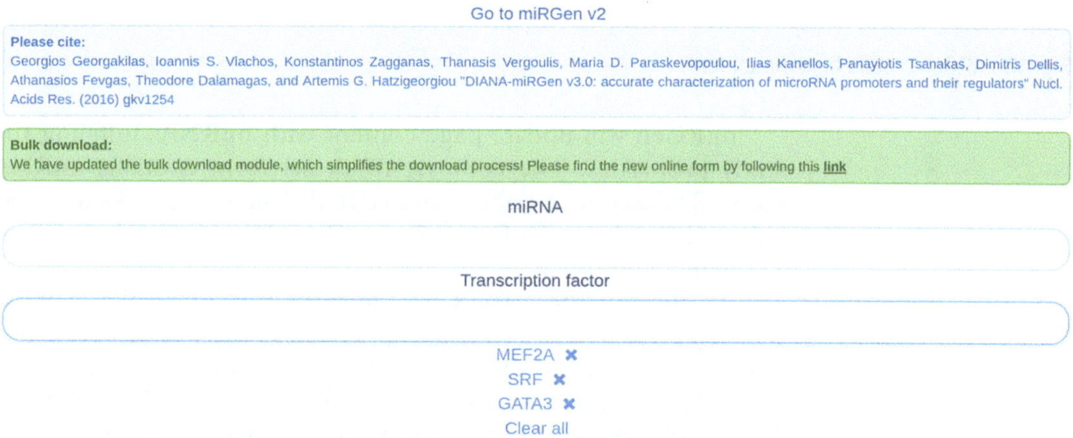

Fig. 8 miRGen initial form. Users have to fill the search fields with the TF or miRNA names of interest

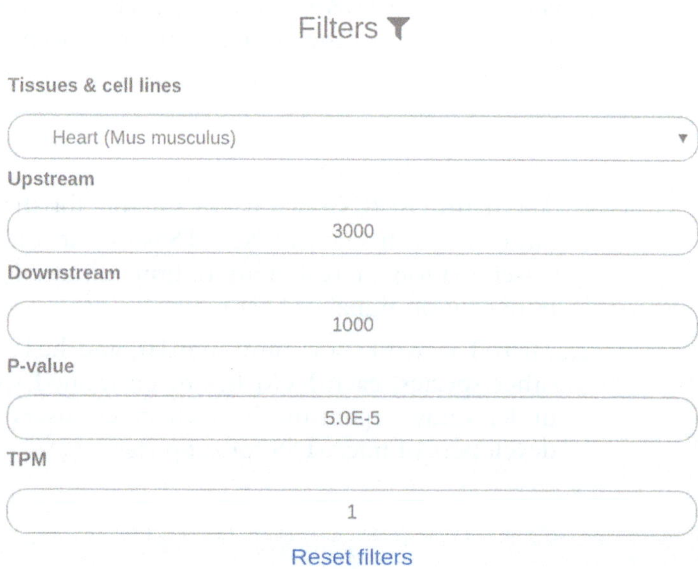

Fig. 9 miRGen options form (left pane)

miRNA name	TSS Coordinates	Tissue & cell line	DIANA Links
mmu-let-7c-1	chr16:77594602-77594603 [+]	Heart (Mus musculus)	mT TB lnE lnP mP
mmu-mir-125b-2	chr16:77594602-77594603 [+]	Heart (Mus musculus)	mT TB lnE lnP mP
mmu-mir-133a-2	chr2:180385585-180385586 [+]	Heart (Mus musculus)	mT TB lnE lnP mP
mmu-mir-15a	chr14:61682345-61682346 [-]	Heart (Mus musculus)	mT TB lnE lnP mP
mmu-mir-16-1	chr14:61682345-61682346 [-]	Heart (Mus musculus)	mT TB lnE lnP mP
mmu-mir-1a-1	chr2:180385585-180385586 [+]	Heart (Mus musculus)	mT TB lnE lnP mP
mmu-mir-23a	chr8:84200977-84200978 [+]	Heart (Mus musculus)	mT TB lnE lnP mP
mmu-mir-24-2	chr8:84200977-84200978 [+]	Heart (Mus musculus)	mT TB lnE lnP mP
mmu-mir-27a	chr8:84200977-84200978 [+]	Heart (Mus musculus)	mT TB lnE lnP mP
mmu-mir-99a	chr16:77594602-77594603 [+]	Heart (Mus musculus)	mT TB lnE lnP mP

Fig. 10 miRGen results in compact form

miRGen will update page content with miRNAs fulfilling the study properties. The compact result set form contains miRNA name, TSS genomic coordinates related tissues, and links to external tools such as TarBase v7, LncBase Experimental, LncBase Predicted, and miRPath v3 (Fig. 10).

By clicking the arrows on the right side of results, a more detailed form of data will come up, enabling users to explore information related to the expression of the TFs found to regulate each miRNA, genomic coordinates of TF binding sites, TF motif logo, and the distance of TFBS from the predicted TSS (Fig. 11).

The study could be continued with other DIANA tools to explore the post-transcriptional regulatory landscape of these miRNAs by querying DIANA-microT, TarBase, and LncBase (This is beyond the scope of the presented use case scenario).

4 Notes

1. The nature of RNA-seq is not suitable for the identification of intragenic/intronic miRNAs TSSs due to the weakness in the dissemination of real transcription initiation events from the transcription signal of host genes.
2. MicroTSS works for mm9, mm10, and hg19. For supporting other species, each SVM has to be trained specifically for the under-study organism. In such cases, users can contact the developers of microTSS for support.

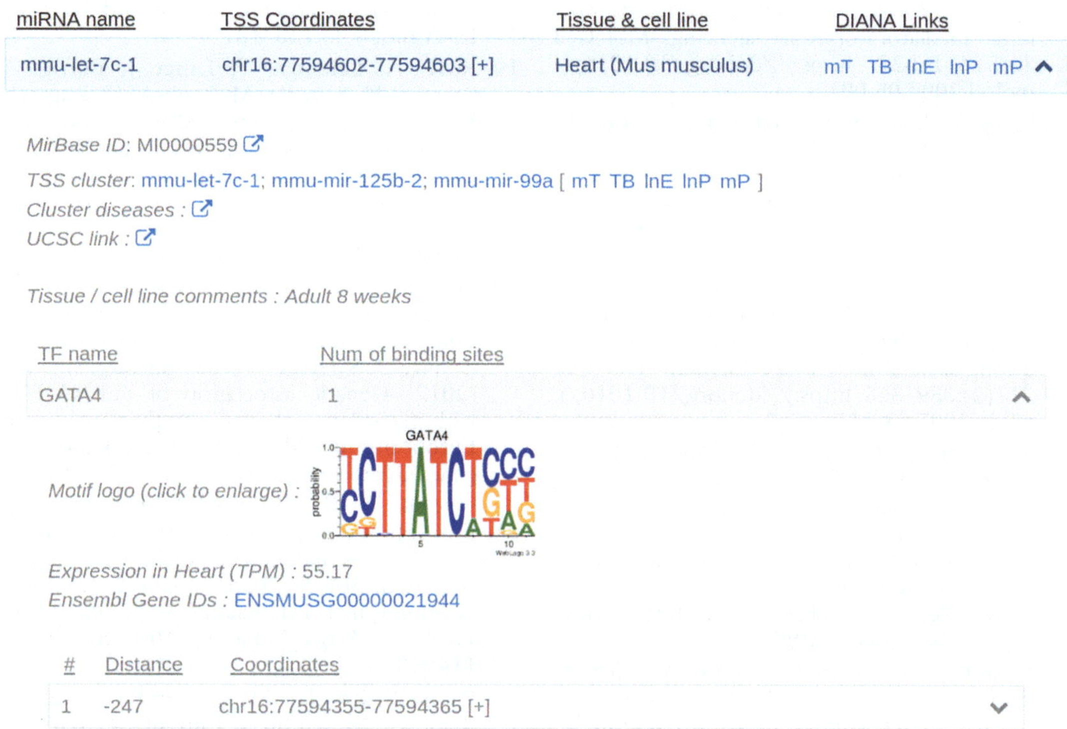

Fig. 11 miRGen results in detailed form

References

1. Lee RC, Feinbaum RL, Ambros V (1993) The C. elegans heterochronic gene lin-4 encodes small RNAs with antisense complementarity to lin-14. Cell 75(5):843–854
2. Georgakilas G, Vlachos IS, Paraskevopoulou MD, Yang P, Zhang Y, Economides AN, Hatzigeorgiou AG (2014) microTSS: accurate microRNA transcription start site identification reveals a significant number of divergent pri-miRNAs. Nat Commun 5:5700. https://doi.org/10.1038/ncomms6700
3. Lee Y, Jeon K, Lee JT, Kim S, Kim VN (2002) MicroRNA maturation: stepwise processing and subcellular localization. EMBO J 21(17):4663–4670
4. Iorio MV, Ferracin M, Liu CG, Veronese A, Spizzo R, Sabbioni S, Magri E, Pedriali M, Fabbri M, Campiglio M, Menard S, Palazzo JP, Rosenberg A, Musiani P, Volinia S, Nenci I, Calin GA, Querzoli P, Negrini M, Croce CM (2005) MicroRNA gene expression deregulation in human breast cancer. Cancer Res 65(16):7065–7070.https://doi.org/10.1158/0008-5472.CAN-05-1783
5. Lee Y, Ahn C, Han J, Choi H, Kim J, Yim J, Lee J, Provost P, Radmark O, Kim S, Kim VN (2003) The nuclear RNase III Drosha initiates microRNA processing. Nature 425(6956): 415–419
6. Denli AM, Tops BB, Plasterk RH, Ketting RF, Hannon GJ (2004) Processing of primary microRNAs by the microprocessor complex. Nature 432(7014):231–235
7. Han J, Lee Y, Yeom KH, Nam JW, Heo I, Rhee JK, Sohn SY, Cho Y, Zhang BT, Kim VN (2006) Molecular basis for the recognition of primary microRNAs by the Drosha-DGCR8 complex. Cell 125(5):887–901. https://doi.org/10.1016/j.cell.2006.03.043
8. Hutvagner G, McLachlan J, Pasquinelli AE, Balint E, Tuschl T, Zamore PD (2001) A cellular function for the RNA-interference enzyme dicer in the maturation of the let-7 small temporal RNA. Science 293(5531):834–838. https://doi.org/10.1126/science.1062961
9. Khvorova A, Reynolds A, Jayasena SD (2003) Functional siRNAs and miRNAs exhibit strand bias. Cell 115(2):209–216

10. Peters L, Meister G (2007) Argonaute proteins: mediators of RNA silencing. Mol Cell 26(5):611–623. https://doi.org/10.1016/j.molcel.2007.05.001
11. Wang Y, Medvid R, Melton C, Jaenisch R, Blelloch R (2007) DGCR8 is essential for microRNA biogenesis and silencing of embryonic stem cell self-renewal. Nat Genet 39(3):380–385. https://doi.org/10.1038/ng1969
12. O'Rourke JR, Georges SA, Seay HR, Tapscott SJ, McManus MT, Goldhamer DJ, Swanson MS, Harfe BD (2007) Essential role for dicer during skeletal muscle development. Dev Biol 311(2):359–368. https://doi.org/10.1016/j.ydbio.2007.08.032
13. Zhu Y, Wang D, Wang F, Li T, Dong L, Liu H, Ma Y, Jiang F, Yin H, Yan W, Luo M, Tang Z, Zhang G, Wang Q, Zhang J, Zhou J, Yu J (2013) A comprehensive analysis of GATA-1-regulated miRNAs reveals miR-23a to be a positive modulator of erythropoiesis. Nucleic Acids Res 41(7):4129–4143. https://doi.org/10.1093/nar/gkt093
14. Calin GA, Dumitru CD, Shimizu M, Bichi R, Zupo S, Noch E, Aldler H, Rattan S, Keating M, Rai K, Rassenti L, Kipps T, Negrini M, Bullrich F, Croce CM (2002) Frequent deletions and down-regulation of micro- RNA genes miR15 and miR16 at 13q14 in chronic lymphocytic leukemia. Proc Natl Acad Sci U S A 99(24):15524–15529. https://doi.org/10.1073/pnas.242606799
15. Yanaihara N, Caplen N, Bowman E, Seike M, Kumamoto K, Yi M, Stephens RM, Okamoto A, Yokota J, Tanaka T, Calin GA, Liu CG, Croce CM, Harris CC (2006) Unique microRNA molecular profiles in lung cancer diagnosis and prognosis. Cancer Cell 9(3):189–198. https://doi.org/10.1016/j.ccr.2006.01.025
16. Shimizu S, Takehara T, Hikita H, Kodama T, Miyagi T, Hosui A, Tatsumi T, Ishida H, Noda T, Nagano H, Doki Y, Mori M, Hayashi N (2010) The let-7 family of microRNAs inhibits Bcl-xL expression and potentiates sorafenib-induced apoptosis in human hepatocellular carcinoma. J Hepatol 52(5):698–704. https://doi.org/10.1016/j.jhep.2009.12.024
17. Kong W, He L, Coppola M, Guo J, Esposito NN, Coppola D, Cheng JQ (2010) MicroRNA-155 regulates cell survival, growth, and chemosensitivity by targeting FOXO3a in breast cancer. J Biol Chem 285(23):17869–17879. https://doi.org/10.1074/jbc.M110.101055
18. Cho WC, Chow AS, Au JS (2011) MiR-145 inhibits cell proliferation of human lung adenocarcinoma by targeting EGFR and NUDT1. RNA Biol 8(1):125–131
19. Keller A, Leidinger P, Lange J, Borries A, Schroers H, Scheffler M, Lenhof HP, Ruprecht K, Meese E (2009) Multiple sclerosis: microRNA expression profiles accurately differentiate patients with relapsing-remitting disease from healthy controls. PLoS One 4(10):e7440. https://doi.org/10.1371/journal.pone.0007440
20. Lofgren SE, Frostegard J, Truedsson L, Pons-Estel BA, D'Alfonso S, Witte T, Lauwerys BR, Endreffy E, Kovacs L, Vasconcelos C, Martins da Silva B, Kozyrev SV, Alarcon-Riquelme ME (2012) Genetic association of miRNA-146a with systemic lupus erythematosus in Europeans through decreased expression of the gene. Genes Immun 13(3):268–274. https://doi.org/10.1038/gene.2011.84
21. Kim J, Inoue K, Ishii J, Vanti WB, Voronov SV, Murchison E, Hannon G, Abeliovich A (2007) A MicroRNA feedback circuit in midbrain dopamine neurons. Science 317(5842):1220–1224. https://doi.org/10.1126/science.1140481
22. Pfeffer S, Zavolan M, Grasser FA, Chien M, Russo JJ, Ju J, John B, Enright AJ, Marks D, Sander C, Tuschl T (2004) Identification of virus-encoded microRNAs. Science 304(5671):734–736. https://doi.org/10.1126/science.1096781
23. Sullivan CS, Grundhoff AT, Tevethia S, Pipas JM, Ganem D (2005) SV40-encoded microRNAs regulate viral gene expression and reduce susceptibility to cytotoxic T cells. Nature 435(7042):682–686. https://doi.org/10.1038/nature03576
24. Andersson MG, Haasnoot PC, Xu N, Berenjian S, Berkhout B, Akusjarvi G (2005) Suppression of RNA interference by adenovirus virus-associated RNA. J Virol 79(15):9556–9565. https://doi.org/10.1128/JVI.79.15.9556-9565.2005
25. Cai X, Lu S, Zhang Z, Gonzalez CM, Damania B, Cullen BR (2005) Kaposi's sarcoma-associated herpesvirus expresses an array of viral microRNAs in latently infected cells. Proc Natl Acad Sci U S A 102(15):5570–5575. https://doi.org/10.1073/pnas.0408192102
26. Zhou X, Ruan J, Wang G, Zhang W (2007) Characterization and identification of microRNA core promoters in four model species. PLoS Comput Biol 3(3):e37. https://doi.org/10.1371/journal.pcbi.0030037
27. Saini HK, Enright AJ, Griffiths-Jones S (2008) Annotation of mammalian primary microRNAs. BMC Genomics 9:564. https://doi.org/10.1186/1471-2164-9-564

28. Saini HK, Griffiths-Jones S, Enright AJ (2007) Genomic analysis of human microRNA transcripts. Proc Natl Acad Sci U S A 104(45):17719–17724. https://doi.org/10.1073/pnas.0703890104
29. Megraw M, Pereira F, Jensen ST, Ohler U, Hatzigeorgiou AG (2009) A transcription factor affinity-based code for mammalian transcription initiation. Genome Res 19(4):644–656. https://doi.org/10.1101/gr.085449.108
30. Barski A, Jothi R, Cuddapah S, Cui K, Roh TY, Schones DE, Zhao K (2009) Chromatin poises miRNA- and protein-coding genes for expression. Genome Res 19(10):1742–1751. https://doi.org/10.1101/gr.090951.109
31. Ozsolak F, Poling LL, Wang Z, Liu H, Liu XS, Roeder RG, Zhang X, Song JS, Fisher DE (2008) Chromatin structure analyses identify miRNA promoters. Genes Dev 22(22):3172–3183. https://doi.org/10.1101/gad.1706508
32. Corcoran DL, Pandit KV, Gordon B, Bhattacharjee A, Kaminski N, Benos PV (2009) Features of mammalian microRNA promoters emerge from polymerase II chromatin immunoprecipitation data. PLoS One 4(4):e5279. https://doi.org/10.1371/journal.pone.0005279
33. Chien CH, Sun YM, Chang WC, Chiang-Hsieh PY, Lee TY, Tsai WC, Horng JT, Tsou AP, Huang HD (2011) Identifying transcriptional start sites of human microRNAs based on high-throughput sequencing data. Nucleic Acids Res 39(21):9345–9356. https://doi.org/10.1093/nar/gkr604
34. Marsico A, Huska MR, Lasserre J, Hu H, Vucicevic D, Musahl A, Orom U, Vingron M (2013) PROmiRNA: a new miRNA promoter recognition method uncovers the complex regulation of intronic miRNAs. Genome Biol 14(8):R84.https://doi.org/10.1186/gb-2013-14-8-r84
35. Economides AN, Frendewey D, Yang P, Dominguez MG, Dore AT, Lobov IB, Persaud T, Rojas J, McClain J, Lengyel P, Droguett G, Chernomorsky R, Stevens S, Auerbach W, Dechiara TM, Pouyemirou W, Cruz JM Jr, Feeley K, Mellis IA, Yasenchack J, Hatsell SJ, Xie L, Latres E, Huang L, Zhang Y, Pefanis E, Skokos D, Deckelbaum RA, Croll SD, Davis S, Valenzuela DM, Gale NW, Murphy AJ, Yancopoulos GD (2013) Conditionals by inversion provide a universal method for the generation of conditional alleles. Proc Natl Acad Sci U S A 110(34):E3179–E3188. https://doi.org/10.1073/pnas.1217812110
36. Megraw M, Sethupathy P, Corda B, Hatzigeorgiou AG (2007) miRGen: a database for the study of animal microRNA genomic organization and function. Nucleic Acids Res 35(Database issue):D149–D155. https://doi.org/10.1093/nar/gkl904
37. Alexiou P, Vergoulis T, Gleditzsch M, Prekas G, Dalamagas T, Megraw M, Grosse I, Sellis T, Hatzigeorgiou AG (2010) miRGen 2.0: a database of microRNA genomic information and regulation. Nucleic Acids Res 38(Database issue):D137–D141. https://doi.org/10.1093/nar/gkp888
38. Yang JH, Li JH, Jiang S, Zhou H, Qu LH (2013) ChIPBase: a database for decoding the transcriptional regulation of long non-coding RNA and microRNA genes from ChIP-Seq data. Nucleic Acids Res 41(Database issue):D177–D187. https://doi.org/10.1093/nar/gks1060
39. Friard O, Re A, Taverna D, De Bortoli M, Cora D (2010) CircuitsDB: a database of mixed microRNA/transcription factor feed-forward regulatory circuits in human and mouse. BMC Bioinformatics 11:435. https://doi.org/10.1186/1471-2105-11-435
40. Wang S, Li W, Lian B, Liu X, Zhang Y, Dai E, Yu X, Meng F, Jiang W, Li X (2015) TMREC: a database of transcription factor and MiRNA regulatory cascades in human diseases. PLoS One 10(5):e0125222. https://doi.org/10.1371/journal.pone.0125222
41. Wang J, Lu M, Qiu C, Cui Q (2010) TransmiR: a transcription factor-microRNA regulation database. Nucleic Acids Res 38(Database issue):D119–D122. https://doi.org/10.1093/nar/gkp803
42. Gillis AJ, Rijlaarsdam MA, Eini R, Dorssers LC, Biermann K, Murray MJ, Nicholson JC, Coleman N, Dieckmann KP, Belge G, Bullerdiek J, Xu T, Bernard N, Looijenga LH (2013) Targeted serum miRNA (TSmiR) test for diagnosis and follow-up of (testicular) germ cell cancer patients: a proof of principle. Mol Oncol 7(6):1083–1092. https://doi.org/10.1016/j.molonc.2013.08.002
43. Georgakilas G, Vlachos IS, Zagganas K, Vergoulis T, Paraskevopoulou MD, Kanellos I, Tsanakas P, Dellis D, Fevgas A, Dalamagas T, Hatzigeorgiou AG (2016) DIANA-miRGen v3.0: accurate characterization of microRNA promoters and their regulators. Nucleic Acids Res 44(D1):D190–D195. https://doi.org/10.1093/nar/gkv1254
44. Vlachos IS, Vergoulis T, Paraskevopoulou MD, Lykokanellos F, Georgakilas G, Georgiou P, Chatzopoulos S, Karagkouni D, Christodoulou

F, Dalamagas T, Hatzigeorgiou AG (2016) DIANA-mirExTra v2.0: uncovering microRNAs and transcription factors with crucial roles in NGS expression data. Nucleic Acids Res 44(W1):W128–W134. https://doi.org/10.1093/nar/gkw455

45. Paraskevopoulou MD, Georgakilas G, Kostoulas N, Vlachos IS, Vergoulis T, Reczko M, Filippidis C, Dalamagas T, Hatzigeorgiou AG (2013) DIANA-microT web server v5.0: service integration into miRNA functional analysis workflows. Nucleic Acids Res 41(Web Server issue):W169–W173. https://doi.org/10.1093/nar/gkt393
46. Karagkouni D, Paraskevopoulou MD, Chatzopoulos S, Vlachos IS, Tastsoglou S, Kanellos I, Papadimitriou D, Kavakiotis I, Maniou S, Skoufos G, Vergoulis T, Dalamagas T, Hatzigeorgiou AG (2017) DIANA-TarBase v8: a decade-long collection of experimentally supported miRNA-gene interactions. Nucleic Acids Res 46(D1):D239–D245. https://doi.org/10.1093/nar/gkx1141
47. Paraskevopoulou MD, Vlachos IS, Karagkouni D, Georgakilas G, Kanellos I, Vergoulis T, Zagganas K, Tsanakas P, Floros E, Dalamagas T, Hatzigeorgiou AG (2016) DIANA-LncBase v2: indexing microRNA targets on non-coding transcripts. Nucleic Acids Res 44(D1):D231–D238. https://doi.org/10.1093/nar/gkv1270
48. Vlachos IS, Zagganas K, Paraskevopoulou MD, Georgakilas G, Karagkouni D, Vergoulis T, Dalamagas T, Hatzigeorgiou AG (2015) DIANA-miRPath v3.0: deciphering microRNA function with experimental support. Nucleic Acids Res 43(W1):W460–W466. https://doi.org/10.1093/nar/gkv403
49. Yue F, Cheng Y, Breschi A, Vierstra J, Wu W, Ryba T, Sandstrom R, Ma Z, Davis C, Pope BD, Shen Y, Pervouchine DD, Djebali S, Thurman RE, Kaul R, Rynes E, Kirilusha A, Marinov GK, Williams BA, Trout D, Amrhein H, Fisher-Aylor K, Antoshechkin I, DeSalvo G, See LH, Fastuca M, Drenkow J, Zaleski C, Dobin A, Prieto P, Lagarde J, Bussotti G, Tanzer A, Denas O, Li K, Bender MA, Zhang M, Byron R, Groudine MT, McCleary D, Pham L, Ye Z, Kuan S, Edsall L, Wu YC, Rasmussen MD, Bansal MS, Kellis M, Keller CA, Morrissey CS, Mishra T, Jain D, Dogan N, Harris RS, Cayting P, Kawli T, Boyle AP, Euskirchen G, Kundaje A, Lin S, Lin Y, Jansen C, Malladi VS, Cline MS, Erickson DT, Kirkup VM, Learned K, Sloan CA, Rosenbloom KR, Lacerda de Sousa B, Beal K, Pignatelli M, Flicek P, Lian J, Kahveci T, Lee D, Kent WJ, Ramalho Santos M, Herrero J, Notredame C, Johnson A, Vong S, Lee K, Bates D, Neri F, Diegel M, Canfield T, Sabo PJ, Wilken MS, Reh TA, Giste E, Shafer A, Kutyavin T, Haugen E, Dunn D, Reynolds AP, Neph S, Humbert R, Hansen RS, De Bruijn M, Selleri L, Rudensky A, Josefowicz S, Samstein R, Eichler EE, Orkin SH, Levasseur D, Papayannopoulou T, Chang KH, Skoultchi A, Gosh S, Disteche C, Treuting P, Wang Y, Weiss MJ, Blobel GA, Cao X, Zhong S, Wang T, Good PJ, Lowdon RF, Adams LB, Zhou XQ, Pazin MJ, Feingold EA, Wold B, Taylor J, Mortazavi A, Weissman SM, Stamatoyannopoulos JA, Snyder MP, Guigo R, Gingeras TR, Gilbert DM, Hardison RC, Beer MA, Ren B, Mouse EC (2014) A comparative encyclopedia of DNA elements in the mouse genome. Nature 515(7527):355–364. https://doi.org/10.1038/nature13992
50. Consortium EP (2012) An integrated encyclopedia of DNA elements in the human genome. Nature 489(7414):57–74. https://doi.org/10.1038/nature11247
51. Heinz S, Benner C, Spann N, Bertolino E, Lin YC, Laslo P, Cheng JX, Murre C, Singh H, Glass CK (2010) Simple combinations of lineage-determining transcription factors prime cis-regulatory elements required for macrophage and B cell identities. Mol Cell 38(4):576–589. https://doi.org/10.1016/j.molcel.2010.05.004
52. Piper J, Elze MC, Cauchy P, Cockerill PN, Bonifer C, Ott S (2013) Wellington: a novel method for the accurate identification of digital genomic footprints from DNase-seq data. Nucleic Acids Res 41(21):e201. https://doi.org/10.1093/nar/gkt850
53. Kohli S, Ahuja S, Rani V (2011) Transcription factors in heart: promising therapeutic targets in cardiac hypertrophy. Curr Cardiol Rev 7(4):262–271
54. He A, Kong SW, Ma Q, Pu WT (2011) Co-occupancy by multiple cardiac transcription factors identifies transcriptional enhancers active in heart. Proc Natl Acad Sci U S A 108(14):5632–5637. https://doi.org/10.1073/pnas.1016959108
55. Coppola A, Romito A, Borel C, Gehrig C, Gagnebin M, Falconnet E, Izzo A, Altucci L, Banfi S, Antonarakis SE, Minchiotti G, Cobellis G (2014) Cardiomyogenesis is controlled by the miR-99a/let-7c cluster and epigenetic modifications. Stem Cell Res 12(2):323–337. https://doi.org/10.1016/j.scr.2013.11.008
56. Wei Y, Peng S, Wu M, Sachidanandam R, Tu Z, Zhang S, Falce C, Sobie EA, Lebeche D, Zhao Y (2014) Multifaceted roles of miR-1s in repressing the fetal gene program in the heart.

Cell Res 24(3):278–292. https://doi.org/10.1038/cr.2014.12

57. Izarra A, Moscoso I, Levent E, Canon S, Cerrada I, Diez-Juan A, Blanca V, Nunez-Gil IJ, Valiente I, Ruiz-Sauri A, Sepulveda P, Tiburcy M, Zimmermann WH, Bernad A (2014) miR-133a enhances the protective capacity of cardiac progenitors cells after myocardial infarction. Stem Cell Rep 3(6):1029–1042. https://doi.org/10.1016/j.stemcr.2014.10.010
58. Wang K, Lin ZQ, Long B, Li JH, Zhou J, Li PF (2012) Cardiac hypertrophy is positively regulated by MicroRNA miR-23a. J Biol Chem 287(1):589–599. https://doi.org/10.1074/jbc.M111.266940
59. Wang J, Huang W, Xu R, Nie Y, Cao X, Meng J, Xu X, Hu S, Zheng Z (2012) MicroRNA-24 regulates cardiac fibrosis after myocardial infarction. J Cell Mol Med 16(9):2150–2160. https://doi.org/10.1111/j.1582-4934.2012.01523.x
60. Chen M, Ma G, Yue Y, Wei Y, Li Q, Tong Z, Zhang L, Miao G, Zhang J (2014) Downregulation of the miR-30 family microRNAs contributes to endoplasmic reticulum stress in cardiac muscle and vascular smooth muscle cells. Int J Cardiol 173(1):65–73. https://doi.org/10.1016/j.ijcard.2014.02.007
61. Roca-Alonso L, Castellano L, Mills A, Dabrowska AF, Sikkel MB, Pellegrino L, Jacob J, Frampton AE, Krell J, Coombes RC, Harding SE, Lyon AR, Stebbing J (2015) Myocardial MiR-30 downregulation triggered by doxorubicin drives alterations in beta-adrenergic signaling and enhances apoptosis. Cell Death Dis 6:e1754. https://doi.org/10.1038/cddis.2015.89

Chapter 3

Metabolic Pulse-Chase RNA Labeling for pri-miRNA Processing Dynamics

Annita Louloupi and Ulf Andersson Vang Ørom

Abstract

miRNA biogenesis is a multistep process starting with the cleavage of the primary miRNA transcript in the nucleus by the microprocessor complex. The pri-miRNA processing kinetics has a high impact on the final regulative role of the mature miRNAs on the expression of their target transcripts. Thus studying the in vivo kinetics of the miRNA biogenesis could give more insights into the contribution of each individual miRNA on regulation of gene expression. Here, we describe step by step a method to determine the processing kinetics of pri-miRNAs in vivo, using a pulse-chase approach that can be used in downstream applications such as qPCR or deep sequencing. We explain in detail the various aspects of this approach that can be applied to different mammalian cell types. The nature of this protocol allows the in vivo study of pri-miRNA processing kinetics in cells treated with different conditions, mutants, and/or cancer cell lines under physiological conditions.

Key words pri-miRNA processing dynamics, BrU pulse-chase labeling

1 Introduction

MiRNAs are small noncoding RNAs that regulate post-transcriptionally the expression of their target genes [1]. The miRNA biogenesis starts with transcription of the pri-miRNA by RNA polymerase II, generating pri-miRNAs up to 30 kb long that are capped and polyadenylated [2]. Pri-miRNAs are then processed in the nucleus by the microprocessor complex comprised of Drosha, an RNase III protein, and DGCR8, a dsRNA-binding protein, that serves as a ruler for Drosha's cleavage site [3, 4]. The microprocessor complex generates the precursor miRNA (pre-miRNA) which is then exported and further processed by Dicer, another RNase III protein, in the cytoplasm to generate the mature miRNA [5]. Many in vitro studies have revealed features, such as sequence motifs around the hairpin loop of the pre-miRNA within the pri-miRNA, that are important for the processing efficiency [6, 7]. The in vivo processing kinetics have so far been challenging to

Ulf Andersson Vang Ørom (ed.), *miRNA Biogenesis: Methods and Protocols*, Methods in Molecular Biology, vol. 1823, https://doi.org/10.1007/978-1-4939-8624-8_3, © Springer Science+Business Media, LLC, part of Springer Nature 2018

study due to the lack of an appropriate method. Recently, we have demonstrated that the steady-state processing efficiency of individual pri-miRNAs within the cell can be characterized by sequencing chromatin-associated RNA. We identified a distinctive microprocessor signature in the transcriptome profile from which we can quantify the efficiency of the endogenous processing event. This study showed that the pri-miRNA processing efficiency is one of the most important factors regulating the levels of the mature miRNA [6, 8].

To further study pri-miRNA processing kinetics in vivo we used a pulse-chase approach that allows the incorporation of uridine analogs, such as 5-bromouridine (BrU), into the newly transcribed RNA. BrU is added to the cell culture media for a 15 min pulse, enough to be taken up by the cells and incorporated into the pool of nucleotide-phosphates from the ribonucleoside salvage pathway where BrU is converted to BrUTP. BrU can be incubated in cell media for hours without any toxic effect making it an ideal RNA labeling uridine analog for in vivo studies. After the pulse, cells are collected on one unique time point or at sequential time points after the removal of the metabolic (BrU) labeling media. The newly transcribed RNA is then immunopurified from the total RNA using an anti-BrdU antibody. The isolated BrU-RNAs can be used directly as a template for cDNA synthesis since BrU does not cause any misincorporation by the reverse transcriptase [9]. To measure pri-miRNA processing kinetics, the BrU pulse is reduced to 15 min and then BrU labeled RNA is chased for 15 min, 30 min, and 1 h. The immunopurified RNA is then subjected to next-generation sequencing or qPCR for further analysis (Fig. 1).

The protocol described above has been used in other studies to answer questions regarding RNA halflife, RNA splicing kinetics and alternative splicing [10]. We have successfully used this approach to study the pri-miRNA processing kinetics [11]. It is a relatively simple and reproducible method that can be used in different cell types under different biological conditions.

2 Materials

2.1 Lab Equipment

1. Cell culture dish 100 × 20 mm.
2. Laminar flow hood.
3. Cell culture incubator.
4. Refrigerated microcentrifuge.
5. 1.5 mL Protein LoBind tubes.
6. 1.5 mL DNA Lobind tubes.
7. 5 mL Eppendorf tubes.
8. Refrigerated Eppendorf 1.5 mL shaker.

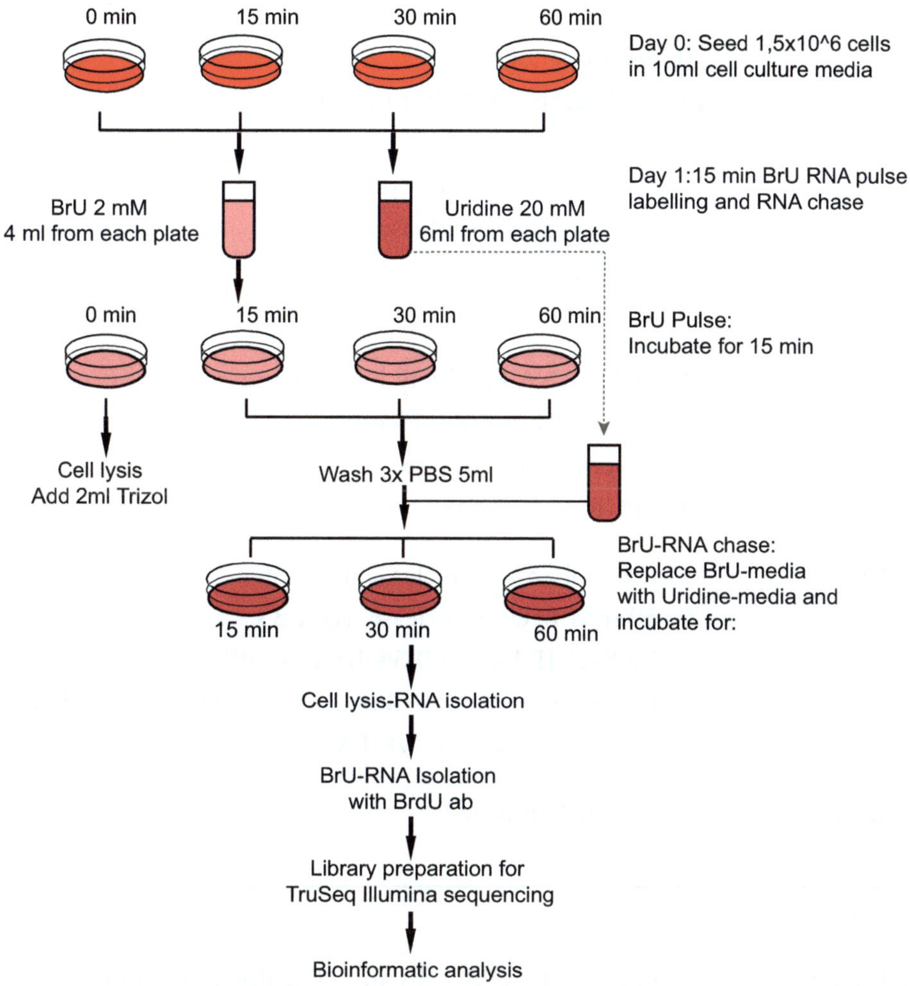

Fig. 1 Metabolic pulse-chase RNA labeling scheme

9. Bioanalyzer.
10. Qubit® Fluorometer.
11. Dynal magnetic separation rack.
12. Tube Rotator.

2.2 Chemicals

1. DMEM, high glucose, no glutamine, no phenol red.
2. Fetal Bovine Serum (FBS).
3. (−)-5-Bromouridine (BrU).
4. Uridine ≥99%.
5. PBS, pH 7.4.
6. TRIzol Reagent.
7. GlycoBlue.
8. Chloroform.

9. Isoamyl alcohol.
10. Isopropanol.
11. BSA, acetylated (20 mg/mL).
12. Ethanol.
13. 3 M sodium acetate (pH 5.2).
14. SUPERase• IN RNase Inhibitor.
15. Distilled water.
16. Protein G Dynabeads.

2.3 Kits

1. Qubit RNA HS Assay Kit-100 assays.
2. Agilent RNA 6000 Pico kit.
3. Reverse Transcriptase.
4. qPCR Master mix.

2.4 Solutions

1. Ethanol 75% in nuclease free water.
2. 50 mM BrU in nuclease free PBS.
3. 5× BrU-IP buffer: 0.5% BSA, 5× PBS in nuclease free water.
4. Elution Buffer: 0.1% BSA and 25 mM bromouridine in PBS.
5. Cell culture media: DMEM supplemented with 10% FBS.

2.5 Antibodies

1. Anti-BrU purified.

3 Methods

3.1 Cell Culture

1. Grow HEK293 cells in cell culture medium under normal growth conditions (typically 37 °C and 5% CO_2) (*see* **Note 1**).
2. The day before BrU labeling seed ~1.5–2 × 10^6 cells per 100 mm plate in 10 mL cell culture media. Cells need to be at 70–80% confluency before the addition of BrU (*see* **Note 2**). Every plate will be for different time points, e.g., for one experiment with four time points (0 h, 15 min chase, 30 min chase, and 1 h chase) seed four different 100 mm plates.

3.2 BrU Labeling

1. Transfer 4 mL medium from each plate to a fresh tube and add BrU to a final concentration of 2 mM and mix. Transfer the rest 6 mL to a second tube and add uridine to a final concentration of 20 mM and mix (*see* **Note 3**).
2. Transfer 4 mL BrU-media back to cells and incubate cells under normal growth conditions (cell culture incubator) for 15 min.
3. After labeling time remove the BrU-medium and for the time point 0 h add directly to the cells 2 mL of Trizol (*see* **Note 4**).

4. For the rest of the time points remove the BrU-media and wash cells twice with 5 mL 1 × PBS.
5. Transfer 6 mL uridine-media back to the plate and incubate cells under normal growth conditions (cell culture incubator) for the different chase time points.
6. Remove the uridine-media and add 2 mL Trizol.
7. Using cell scraper, gently scrape the cells off the bottom and transfer the lysed cells into a fresh 5 mL tube.
8. Lyse cells by pipetting up and down several times.
9. Incubate the tubes for 15 min on ice to permit complete dissociation of nucleoprotein complexes.
10. Store samples at −80 °C if not isolating RNA immediately.

3.3 Isolation of RNA

1. In each tube, add 0.4 mL of 24:1 chloroform:Isoamyl alcohol (0.2 mL per 1 mL Trizol) and vortex for 10 s.
2. Centrifuge tubes at 4 °C for 20 min at full speed.
3. Transfer the upper aqueous layer to a new tube and perform another round of 24:1 chloroform:Isoamyl alcohol extraction to remove any traces of phenol (*see* **Note 5**).
4. Transfer the upper aqueous layer to a new tube and add 1 mL isopropanol (0.5 mL per 1 mL Trizol).
5. Mix the solution well and allow RNA to precipitate for 0.5–1 h at −20 °C.
6. Collect the precipitated RNA by centrifugation at full speed for 30 min at 4 °C.
7. Remove supernatant and wash the pellet twice with 75% ethanol, centrifuge again, and remove any remaining ethanol.
8. Resuspend the pellet in 80 μL RNase-free water.
9. Store samples at −80 °C unless immediately proceeding to BrU-IP.

3.4 BrU-RNA Isolation

1. Transfer 40 μL of anti-mouse IgG magnetic Dynabeads per sample (100 μg −150 μg BrU labeled RNA) to a 1.5 mL microfuge Protein LoBind tube.
2. Capture beads with magnetic stand and remove storage buffer.
3. Add 800 μL 1x BrU-IP buffer and resuspend beads. Perform a quick spin, capture beads on the magnetic stand, and aspirate supernatant. Repeat two more times for a total of three washes.
4. After final wash, resuspend the beads in 1x BrU-IP buffer supplemented with RNase Inhibitor 1:2000.
5. For every sample, add 5 μg BrdU antibody (*see* **Note 6**).

6. Incubate for 1 h at room temperature with gentle rotation.
7. Wash beads three times as above and after final wash resuspend the beads in 100 μL 1× BrU-IP buffer supplemented with RNase Inhibitor 1:2000.
8. In 80 μL of 100–150 μg BrU labeled RNA add 20 μL 5× BrU-IP buffer and the magnetic beads conjugated with the BrdU. Add 1× BrU-IP buffer to have a final IP volume of 0.8 mL (*see* **Note 7**).
9. Incubate for 60 min at room temperature with gentle rotation in Protein LoBind tubes.
10. Wash beads with BrU-IP 1× buffer for 5 min on the rotator. Perform a quick spin, capture beads on the magnetic stand, and aspirate supernatant. Do three additional washes with BrU-IP 1× buffer. At the last wash, change the tubes from protein low binding tubes to DNA LoBind tubes and make sure to completely remove the final wash.
11. Add 200 μL of elution buffer (0.1% BSA and 25 mM bromouridine in PBS).
12. Incubate for 1 h with continuous shaking (1100 rpm) at 4 °C.
13. Spin down the beads and carefully transfer the supernatant to a clean tube.
14. Take care not to take over any beads as it will increase the background (*see* **Note 8**).
15. Add 1/10 volumes of 3 M sodium acetate (pH 5.2), 2.5–3 volumes of 100% ethanol, and 1 μL of GlycoBlue.
16. Vortex the contents for 10 s and allow RNA to precipitate at −80 °C overnight.
17. Centrifuge the tubes at full speed for 30 min at 4 °C. Discard the supernatant.
18. Wash the pellet twice with 75% ethanol, centrifuge again, and remove any remaining ethanol.
19. Carefully aspirate the supernatant and let the pellet air-dry.
20. Resuspend the pellet in 15 μL RNase-free water
21. Perform Qubit Fluorometric Quantitation with Qubit RNA HS Assay Kit according to the manufacturer's instructions to measure the RNA concentration. Expected IP recovery is 0.09–0.16% of input.
22. Check quality of BrU-RNA using Agilent RNA 6000 Pico kit following manufacturer's instructions (*see* **Note 9**).

3.5 Library Preparation and Massively Parallel Sequencing

1. Use 100 ng of BrU-RNA to continue for library preparation following the TruSeq Stranded mRNA Library Preparation Kit instructions with some modifications.

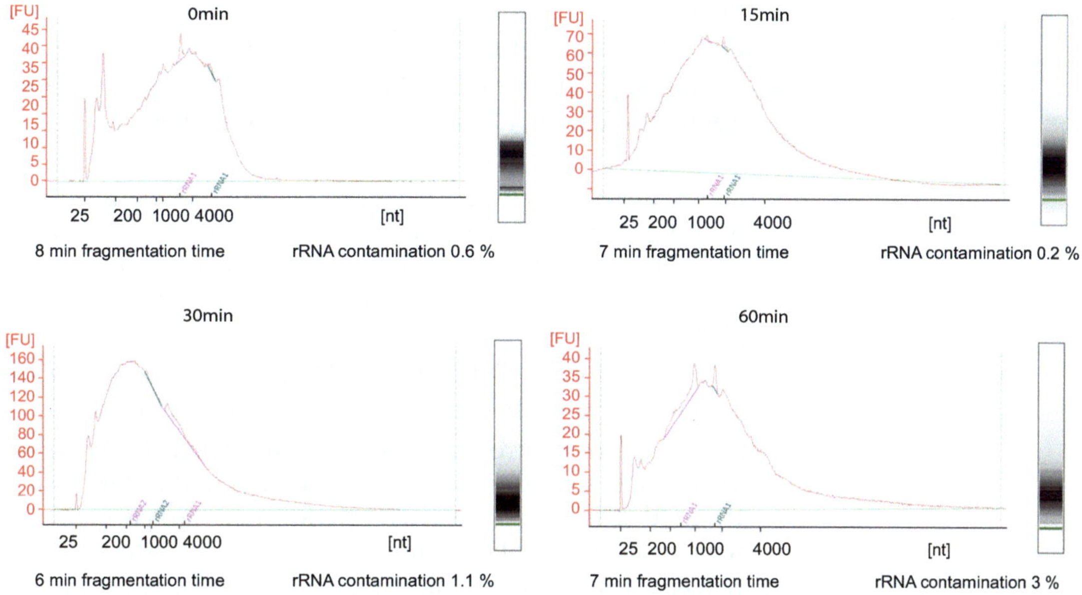

Fig. 2 Bioanalyzer results. The indicated RNA fragmentation time is according to TruSeq Stranded mRNA library prep kit protocol

2. According to the Bioanalyzer results, RNA has to be fragmented in different incubation times at 94 °C (Fig. 2). When using Illumina HiSeq 2500 for sequencing, aim at obtaining around 200 M reads per sample.

3.6 Primer Design for RT-qPCR to Detect the Unprocessed and Processed Pri-miRNA Transcript

1. Use UCSC server to get the sequence for primary miRNA of your interest.
2. Detect the mature miRNA and design one of your primers to anneal at the cleavage site or design the primer to produce an amplicon surrounding Drosha's cleavage site (*see* **Note 10**).

3.7 Transcript Quantification by Reverse Transcription Followed by Quantitative PCR

1. Prepare RT reactions using reverse transcriptase. Use BrU-RNA 10–100 ng as input for reverse transcription and perform the reaction according to manufacturer's instructions. The cDNA can be quantified on a real-time PCR system (*see* **Note 11**).

4 Notes

1. Different cell types can be used such as HeLa or MCF7 breast cancer cell lines.
2. Cells need to be 70–80% confluent to ensure physiological transcription and RNA labeling.

3. Prepare a master mix of conditional medium with BrU and uridine to ensure fast processing of multiple plates.
4. In this step, PBS can be added to the cells and transfer the plates to a fume hood where Trizol can be added.
5. To obtain very pure RNA from Trizol samples in this step the upper aqueous layer can be extracted a second time using acidic phenol pH 4.5–5 1:1 volume followed by another extraction with 24:1 chloroform:Isoamyl alcohol to remove any traces of phenol. The DNA remains in the organic phase in the acidic phenol extraction.
6. If you have more BrU labeled RNA as input, scale up the amount of antibody, dynabeads, and buffer accordingly.
7. As a negative control take along an unlabeled RNA sample. This sample should NOT recover any RNA in the elution step. If yes, that means you need to perform additional washes to remove any unspecific binding of unlabeled RNA molecules.
8. In order to avoid beads contaminations perform another round of quick spin, and carefully transfer the supernatant to a new tube.
9. The rRNA % detection in the bioanalyzer should be from 0 to 3%.
10. The 5′ end of 5p miRNA and 3′ end of 3p miRNA are generally assumed as DROSHA cleavage site.
11. Any cDNA and RT-qPCR kit can be used according to your preference and the starting RNA amount for the cDNA synthesis can be optimized accordingly.

References

1. Kim VN, Han J, Siomi MC (2009) Biogenesis of small RNAs in animals. Nat Rev Mol Cell Biol 10(2):126–139. https://doi.org/10.1038/nrm2632
2. Lee Y, Kim M, Han J, Yeom KH, Lee S, Baek SH, Kim VN (2004) MicroRNA genes are transcribed by RNA polymerase II. EMBO J 23(20):4051–4060. https://doi.org/10.1038/sj.emboj.7600385
3. Han J, Lee Y, Yeom K-H, Kim Y-K, Jin H, Kim VN (2004) The Drosha-DGCR8 complex in primary microRNA processing. Genes Dev 18(24):3016–3027. https://doi.org/10.1101/gad.1262504
4. Gregory RI, Yan KP, Amuthan G, Chendrimada T, Doratotaj B, Cooch N, Shiekhattar R (2004) The microprocessor complex mediates the genesis of microRNAs. Nature 432(7014):235–240
5. Gregory RI, Chendrimada TP, Cooch N, Shiekhattar R (2005) Human RISC couples microRNA biogenesis and posttranscriptional gene silencing. Cell 123(4):631–640. https://doi.org/10.1016/j.cell.2005.10.022
6. Conrad T, Marsico A, Gehre M, Orom UA (2014) Microprocessor activity controls differential miRNA biogenesis in vivo. Cell Rep 9(2):542–554. https://doi.org/10.1016/j.celrep.2014.09.007
7. Auyeung VC, Ulitsky I, McGeary SE, Bartel DP (2013) Beyond secondary structure: primary-sequence determinants license pri-miRNA hairpins for processing. Cell 152(4):844–858. https://doi.org/10.1016/j.cell.2013.01.031
8. Conrad T, Ørom UA (2017) Cellular fractionation and isolation of chromatin-associated RNA. In: Ørom UA (ed) Enhancer RNAs: methods and protocols. Springer, New York,

NY, pp 1–9. https://doi.org/10.1007/978-1-4939-4035-6_1

9. Tani H, Akimitsu N (2012) Genome-wide technology for determining RNA stability in mammalian cells. RNA Biol 9(10):1233–1238. https://doi.org/10.4161/rna.22036
10. Paulsen MT, Veloso A, Prasad J, Bedi K, Ljungman EA, Tsan YC, Chang CW, Tarrier B, Washburn JG, Lyons R, Robinson DR, Kumar-Sinha C, Wilson TE, Ljungman M (2013) Coordinated regulation of synthesis and stability of RNA during the acute TNF-induced proinflammatory response. Proc Natl Acad Sci USA 110(6):2240–2245. https://doi.org/10.1073/pnas.1219192110
11. Louloupi A, Ntini E, Liz J, Ørom UA (2017) Microprocessor dynamics shows co- and post-transcriptional processing of pri-miRNAs. RNA 23(6):892–898. https://doi.org/10.1261/rna.060715.117

Chapter 4

In Vitro System for Coupling RNAP II Transcription to Primary microRNA Processing and a Three-Way System for RNAP II Transcription/Splicing/microRNA Processing

Shanye Yin, Alexander Iocolano, Yong Yu, Jaya Gangopadhyay, and Robin Reed

Abstract

In the genome, primary microRNAs (pri-miRNAs) are encoded either as independent transcriptional units with their own promoters (intergenic miRNAs) or within the introns of other genes (intronic miRNAs). Here, we report two methods, one that we established for coupled RNAP II transcription and pri-miRNA processing and the other that is a three-way system for RNAP II transcription, pri-miRNA processing, and pre-mRNA splicing. In these systems, CMV-DNA constructs encoding the processing substrates are incubated in HeLa cell nuclear extracts in the presence of ^{32}P-UTP to generate the nascent RNAP II transcripts, which are processed efficiently by the endogenous RNA processing machineries in nuclear extracts.

Key words Primary microRNA processing, RNAP II transcription, Pre-mRNA splicing, Coupled steps in gene expression

1 Introduction

It is now well established that RNAP II transcription is functionally coupled to numerous steps in gene expression, including capping, splicing, polyadenylation, pri-miRNA processing, non-sense mediated decay, and mRNA export [1–13]. This coupling potently increases the rate and efficiency of the steps in gene expression and likely enhances the overall fidelity and efficiency of gene expression. MiRNAs are small, non-coding RNAs with critical roles in post-transcriptional control of gene expression [14–17]. In mammalian cells, most miRNAs are transcribed by RNAP II to generate primary miRNAs (pri-miRNAs) transcripts [14]. The majority of pri-miRNAs are found within introns of other genes, and these intronic miRNAs are either processed directly from the intron or the intronic miRNAs have their own promoters. In some cases, pri-miRNAs are encoded as individual transcription units, and these intergenic

Ulf Andersson Vang Ørom (ed.), *miRNA Biogenesis: Methods and Protocols*, Methods in Molecular Biology, vol. 1823,
https://doi.org/10.1007/978-1-4939-8624-8_4, © Springer Science+Business Media, LLC, part of Springer Nature 2018

miRNAs are transcribed by their own promoter. Pri-miRNAs are first processed in the nucleus by the microprocessor complex (Drosha and DGCR8) into ~70 nt hairpin precursor miRNAs (pre-miRNAs), which are then exported to the cytoplasm and further processed by Dicer to generate mature miRNAs [18–22]. The steps of the pri-miRNA processing reaction were first elucidated using in vitro systems in which naked pri-miRNA transcripts generated by T7 RNA polymerase were processed in whole cell lysates or by purified Drosha/DGCR8 [20–22]. In vivo studies subsequently reported that pri-miRNA processing is coupled to RNAP II transcription [23]. To understand the mechanisms underlying this coupling, we previously developed an in vitro system in which RNAP II transcription is functionally coupled to pri-miRNA processing [12]. In addition, we have also developed a three-way system for RNAP II transcription/pri-miRNA processing/splicing [24]. In the first system, pri-miRNA is transcribed by RNAP II and processed in HeLa cell nuclear extracts [12]. In the second, a miRNA is present in the intron of a splicing substrate. The pre-mRNA containing the intronic miRNA is transcribed by RNAP II and then undergoes both splicing and pri-miRNA processing [24]. These systems provide useful tools for investigating the mechanisms involved in linking RNAP II transcription, splicing, and pri-miRNA processing.

2 Materials

All solutions are prepared using analytical grade reagents and ultrapure water (Milli-Q purified deionized water at a sensitivity of 18 MΩ cm at 25 °C). Storage temperature of each reagent is listed below.

2.1 Preparation of CMV-DNA Templates from Plasmids

1. CMV-DNA plasmids encoding pri-let-7a-1 (CMV-let-7a) or FTZ pre-mRNA containing pri-let-7a in the intron (CMV-FTZ-let) should be stored at −20 °C in 1× TE buffer (Tris–HCl, pH 8.0, 1 mM EDTA). Dilute the plasmids to a working concentration of 5 ng/μL before use.
2. CMV-DNA templates are generated by PCR. The PCR primers are CMV_F: 5′-TGGAGGT CGCTGAGTAGTGC-3′ and CMV_R: 5′-TAGAAGGCACAGTCGAGG-3′. Make 100 μL aliquots at a final concentration of 10 μM and store at −20 °C.
3. 10 mM dNTPs. Store at −20 °C.
4. 50 mM $MgSO_4$. Store at −20 °C.
5. Platinum Taq HiFi and 10× HiFi Buffer provided by supplier (Invitrogen). Store at −20 °C.
6. 10× TBE: Combine 432 g Tris Base, 220 g boric acid, and 37.2 g EDTA. Add water to a final volume of 4 L. Store at room temperature.

7. 10 mg/mL ethidium bromide in H_2O. Ethidium bromide is light sensitive, so store in a dark container.
8. Agarose HS standard/high melt. Store at room temperature.
9. 3 M sodium acetate. Store at room temperature.
10. 100% and 70% ethanol diluted from 200 proof pure ethanol. Store at room temperature.
11. Phenol/chloroform, pH 7.9. Store at 4 °C.
12. 1 kb DNA ladder marker. Store at −20 °C.

2.2 Coupled Transcription/Pri-miRNA Processing Reaction

1. 12.5 mM ATP. Filter, make 100 μL aliquots, and store at −20 °C.
2. 0.5 M Creatine Phosphate di-Tris salt: Filter, prepare 100 μL aliquots, and store at −20 °C.
3. 80 mM $MgCl_2$: Filter, prepare 100 μL aliquots, and store at −20 °C.
4. CMV-DNA templates: Make 50 μL aliquots at 200 ng/μL. Store at −20 °C.
5. ^{32}P-UTP (800 Ci/mmol, 250 μCi). Store at 4 °C.
6. HeLa cell nuclear extract (*see* **Note 1**). Store at −80 °C.
7. α-Amanitin: Dilute to 10 ng/μL with water from 1 mg/mL stock. Store at −20 °C. Prepare 100 μL aliquots, and use a fresh aliquot for each experiment.
8. 20 mg/mL Glycogen (Roche). Store at −20 °C.
9. Formamide gel-loading dye: Combine 16 mL formamide, 0.4 mL 0.5 M EDTA, 0.8 mL 2.5% xylene cyanol, and 0.8 mL 2.5% bromophenol blue. Mix well and make 1 mL aliquots. Store at −20 °C.

2.3 Denaturing Polyacrylamide Gel

1. Denaturing gel solution (15%): Mix 215 g urea with 50 mL of 10× TBE and 187.5 mL 40% acrylamide–bisacrylamide (w/v 29:1). Bring the volume up to 500 mL with water. Filter and store at 4 °C.
2. N,N,N′,N′-Tetramethylethylenediamine (TEMED). Store at 4 °C.
3. 10% ammonium persulfate (APS). Store at 4 °C.
4. Model V16 polyacrylamide gel electrophoresis system, PROTEAN II xi spacers 0.5 mm, PROTEAN II xi comb (Bio-Rad).
5. Gel-loading tips (Bioexpress): Flat orifice, 83 mm × 0.33 mm diameter.
6. Whatman paper 3 MM Chr.
7. Gel dryer vacuum system.

8. PhosphorImager cassette: Mounted General Purpose, 20 cm × 25 cm, screen and cassette or equivalent.
9. PhosphorImager.

3 Methods

Carry out all procedures on ice unless otherwise specified.

3.1 Preparation of CMV-DNA Templates

1. PCR reaction: Mix 2 μL of the CMV-DNA plasmid (CMV-let-7a or CMV-FTZ-let) (*see* **Note 1**), 2 μL 10 mM dNTPs, 2 μL 50 mM $MgSO_4$, 5 μL 10× HiFi Buffer, 5.0 μL of each primer, 28.6 μL of Milli-Q water, and 0.4 μL Platinum Taq HiFi. Start the PCR reaction at 94 °C for 5 min followed by 32 cycles at 94 °C for 30 s, 55 °C for 30 s, and 68 °C for 2 min.
2. When the PCR reaction is finished, bring the volume up to 150 μL with 1× TE buffer and check the CMV-DNA template by running a small aliquot (1 μL) on a mini-agarose gel next to the 1 kb DNA size marker.
3. Ethanol precipitate the CMV-DNA template. To do this, add 2 μL of glycogen; mix well. Add 1/10 volume 3 M sodium acetate; mix well. Add three volumes of 100% ethanol. Centrifuge at 16,000 × *g* for 15 min to pellet the DNA. Remove the supernatant, without disturbing the pellet. Wash once with 1 mL 70% ethanol. Air-dry the pellet and dissolve it in 100 μL of water.
4. Determine the concentration of the CMV-DNA template using a NanoDrop™ spectrophotometer or a UV spectrophotometer. Confirm the size and quality of the DNA by running 2 μL on an agarose gel next to the 1 kb DNA size marker.

3.2 Coupled RNAP II Transcription/Pri-miRNA Processing System

1. Preheat two water baths, one at 30 °C and another at 37 °C.
2. For assembly of pre-initiation complexes (PICs), incubate the CMV-let-7a DNA template (4 μLs containing 800 ng DNA) with 1 μL $MgCl_2$ (80 mM), 3 μL 15% polyvinyl alcohol (PVA), and 15 μL HeLa nuclear extract (*see* **Note 2**) at 30 °C for 20 min.
3. To initiate RNAP II transcription, add 1 μL ^{32}P-UTP (800 Ci/mmol; Perkin Elmer Life Sciences), 1 μL ATP (12.5 mM), and 1 μL Creatine Phosphate di-Tris salt (500 mM) to the reaction mixture.
4. After transcription for 5 min at 30 °C, add 1 μL of α-amanitin (10 ng/μL) per 25 μL reaction mixture, and mix gently.
5. Place 5 μL of the reaction mixtures in a new tube and add 18.2 μL fresh nuclear extract and 1.8 μL 80 mM $MgCl_2$ to allow pri-miRNA processing at 37 °C for specified times indicated in Fig. 1.

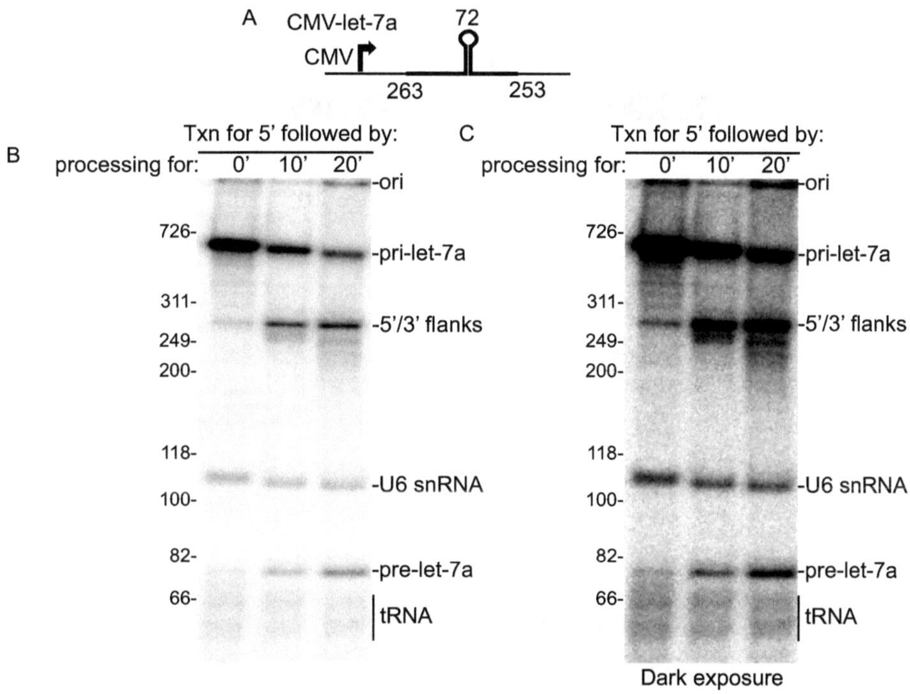

Fig. 1 Coupled RNAP II transcription/pri-miRNA processing in vitro. (**a**) Structure of the CMV-let-7a DNA template. The sizes of 5′ flanking region, 3′ flanking region, and pre-let-7a are indicated. The thick line indicates the natural pri-miRNA sequences and the thin lines indicate the vector sequences. (**b**) CMV-let-7a DNA template was incubated in nuclear extract for 20 min to assemble a pre-initiation complex followed by addition of ^{32}P-UTP, ATP, and Creatine Phosphate and continued incubation for 5 min. α-Amanitin was added to stop transcription and incubation was continued for the indicated times. Pri-miRNAs, pre-miRNAs, and 5′/3′ flanks are indicated. The endogenous U6 snRNA and tRNA in the extract are labeled by ^{32}P-UTP [26]. Markers (in base pairs) are indicated, and Ori indicates the gel origin. (**c**) A dark exposure of (**b**)

6. Remove a 4 μL aliquot of the reaction mixture at each time point and transfer it to a microfuge tube containing 1 mL TRIzol (Invitrogen) (*see* **Note 3**).
7. Repeat **step 6** for all subsequent time points.
8. Add 200 μL of chloroform pH 6.6 to each sample and mix well by vortexing. Centrifuge for 15 min at 16,000 × *g*.
9. Transfer 400 μL of the aqueous phase to a new tube containing 2 μL of glycogen. Mix by vortexing.
10. Add 500 μL of 100% isopropanol and mix by vortexing.
11. Spin at 16,000 × *g* for 15 min. The RNA pellet should look white/translucent.
12. Carefully remove the supernatant using a pipet.
13. Do a quick spin at 16,000 × *g* and remove any remaining supernatant using a P200 pipetman (*see* **Note 4**).
14. Resuspend the pellet in 15 μL formamide gel-loading dye. Heat samples in hot water (75–90 °C) for 10 min.

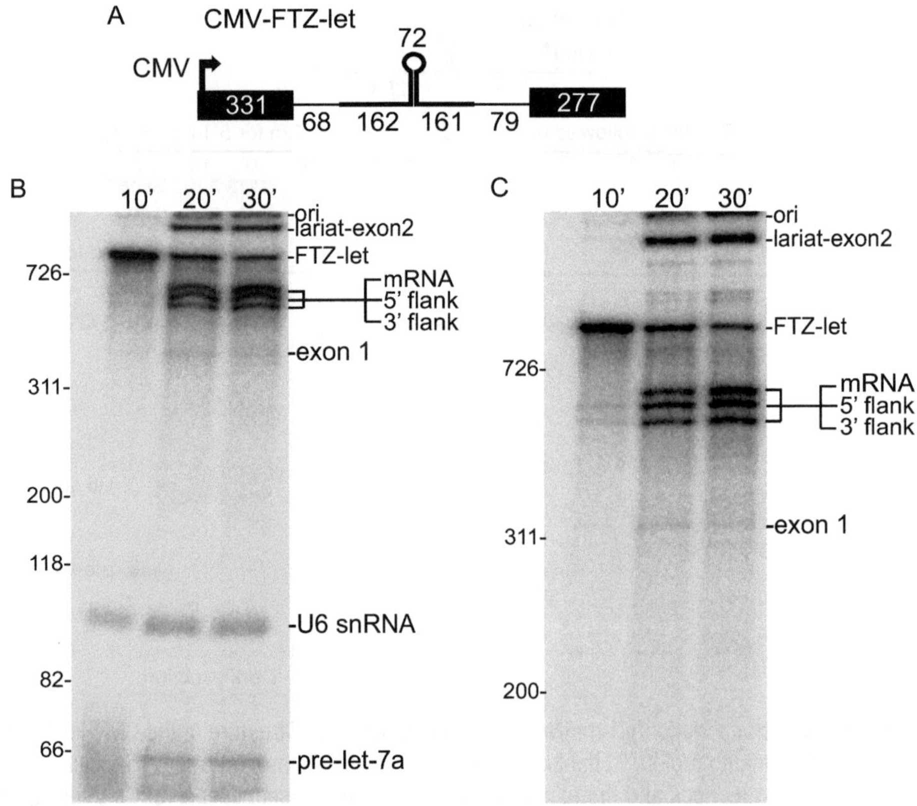

Fig. 2 Coupled RNAP II transcription/splicing/pri-miRNA processing in vitro. (**a**) Structure of the CMV-FTZ-let DNA template. The sizes of FTZ pre-mRNA exons, intron, 5′ flanking region, 3′ flanking region, and pre-let-7a are indicated. The thick line indicates pri-miRNA sequences and the thin lines indicate intron sequences. (**b**) CMV-FTZ-let DNA template was incubated in nuclear extract for 20 min to assemble a pre-initiation complex followed by addition of ^{32}P-UTP, ATP, and Creatine Phosphate and continued incubation for 10 min. α-Amanitin was added to stop transcription and incubation was continued for the indicated times. Full length transcript, spliced mRNA, exon1, pre-miRNAs, and 5′/3′ flanks are indicated. Markers (in base pairs) are indicated and Ori indicates the gel origin. Samples were run for 1 h. (**c**) Same as (**b**), except that samples were run for 2 h to obtain a better separation of pre-mRNA and 5′/3′ flanks

15. Prepare a gel by combining 15 mL of 8% denaturing gel solution with 15 μL TEMED and 150 μL 10% APS. Pour the gel and wait 10 min until it is polymerized.
16. Load 7.5 μL of each sample on the denaturing gel. Run the gel at 20 mAmps, constant current for 45–60 min.
17. Transfer the gel to Whatman paper and dry the gel on a gel dryer for 30 min at 80 °C.
18. Put the gel in a PhosphorImager cassette, expose 1–12 h, and scan the gel using a PhosphorImager.

3.3 Coupled RNAP II Transcription/Splicing/Pri-miRNA Processing System

1. Preheat a water bath to 30 °C.
2. For assembly of pre-initiation complexes (PICs), incubate the CMV-FTZ-let DNA template (4 μLs containing 800 ng DNA) with 1 μL $MgCl_2$ (80 mM), 3 μL 15% polyvinyl alcohol (PVA), and 15 μL HeLa nuclear extract at 30 °C for 20 min.
3. To initiate RNAP II transcription, add 1 μL ^{32}P-UTP (800 Ci/mmol; Perkin Elmer Life Sciences), 1 μL ATP (12.5 mM), and 1 μL Creatine phosphate di-Tris salt (500 mM) to the reaction mixture.
4. After transcription for 10 min at 30 °C, add 1 μL of 10 ng/μL α-amanitin and 1 μL of $MgCl_2$ (80 mM) per 25 μL reaction mixture, and mix gently.
5. Incubate the reaction mixtures at 30 °C for specified times indicated in Fig. 2.
6. Follow **steps 6–18** in Subheading 3.2. For **step 16**, run the gel at 20 mAmps, constant current for 60 or 120 min.

4 Notes

1. In some preparations of nuclear extract, RNAP II transcribes end to end in a promoter-independent manner. If you are using your own DNA template, use the smallest possible PCR fragment that contains your sequence of interest to avoid large end-to-end transcription products.
2. HeLa nuclear extract was prepared as described [25] except that the final centrifugation following dialysis was omitted, which results in increased efficiency of RNAP II transcription.
3. Samples in TRIzol are stable at room temperature and can be stored until all of the time points have been collected. At this step, samples can also be stored at 4 °C overnight and processed later.
4. Ensure that the pellets are dry before adding the formamide loading dye because any remaining ethanol will add to the volume of the sample as well as distort migration of the bands on the gel.

Acknowledgments

We are grateful to members of the Reed lab for comments on the manuscript. This work was supported by an NIH grant GM122524 to RR.

References

1. Greenleaf AL (1993) Positive patches and negative noodles: linking RNA processing to transcription? Trends Biochem Sci 18(4):117–119. https://doi.org/10.1016/(93)90016-g
2. Maniatis T, Reed R (2002) An extensive network of coupling among gene expression machines. Nature 416(6880):499–506
3. Prasanth KV, Spector DL (2007) Eukaryotic regulatory RNAs: an answer to the 'genome complexity' conundrum. Genes Dev 21(1):11–42. https://doi.org/10.1101/gad.1484207
4. Buratowski S (2009) Progression through the RNA polymerase II CTD cycle. Mol Cell 36(4):541–546. https://doi.org/10.1016/j.molcel.2009.10.019
5. Margueron R, Reinberg D (2010) Chromatin structure and the inheritance of epigenetic information. Nat Rev Genet 11(4):285–296. https://doi.org/10.1038/nrg2752
6. Proudfoot NJ (2011) Ending the message: poly(A) signals then and now. Genes Dev 25(17):1770–1782. https://doi.org/10.1101/gad.17268411
7. Hsin JP, Manley JL (2012) The RNA polymerase II CTD coordinates transcription and RNA processing. Genes Dev 26(19):2119–2137. https://doi.org/10.1101/gad.200303.112
8. Gomez Acuna LI, Fiszbein A, Allo M, Schor IE, Kornblihtt AR (2013) Connections between chromatin signatures and splicing. Wiley Interdiscip Rev RNA 4(1):77–91. https://doi.org/10.1002/wrna.1142
9. Brugiolo M, Herzel L, Neugebauer KM (2013) Counting on co-transcriptional splicing. F1000Prime Rep 5:9. https://doi.org/10.12703/P5-9
10. Muller-McNicoll M, Neugebauer KM (2013) How cells get the message: dynamic assembly and function of mRNA-protein complexes. Nat Rev Genet 14(4):275–287. https://doi.org/10.1038/nrg3434
11. Bentley DL (2014) Coupling mRNA processing with transcription in time and space. Nat Rev Genet 15(3):163–175. https://doi.org/10.1038/nrg3662
12. Yin S, Yu Y, Reed R (2015) Primary microRNA processing is functionally coupled to RNAP II transcription in vitro. Sci Rep 5:11992. https://doi.org/10.1038/srep11992
13. Reed R, Hurt E (2002) A conserved mRNA export machinery coupled to pre-mRNA splicing. Cell 108(4):523–531
14. Bartel DP (2004) MicroRNAs: genomics, biogenesis, mechanism, and function. Cell 116(2):281–297
15. Bartel DP, Chen C-Z (2004) Opinion: micromanagers of gene expression: the potentially widespread influence of metazoan microRNAs. Nat Rev Genet 5(5):396–400. https://doi.org/10.1038/nrg1328
16. Ambros V (2004) The functions of animal microRNAs. Nature 431(7006):350–355. https://doi.org/10.1038/nature02871
17. Lee RC, Feinbaum RL, Ambros V (1993) The C. elegans heterochronic gene lin-4 encodes small RNAs with antisense complementarity to lin-14. Cell 75(5):843–854
18. Denli AM, Tops BBJ, Plasterk RHA, Ketting RF, Hannon GJ (2004) Processing of primary microRNAs by the microprocessor complex. Nature 432(7014):231–235. https://doi.org/10.1038/nature03049
19. Gregory RI, K-p Y, Amuthan G, Chendrimada T, Doratotaj B, Cooch N, Shiekhattar R (2004) The microprocessor complex mediates the genesis of microRNAs. Nature 432(7014):235–240. https://doi.org/10.1038/nature03120
20. Lee Y, Ahn C, Han J, Choi H, Kim J, Yim J, Lee J, Provost P, Rådmark O, Kim S, Kim VN (2003) The nuclear RNase III Drosha initiates microRNA processing. Nature 425(6956):415–419. https://doi.org/10.1038/nature01957
21. Lund E (2004) Nuclear export of MicroRNA precursors. Science 303(5654):95–98. https://doi.org/10.1126/science.1090599
22. Meister G, Tuschl T (2004) Mechanisms of gene silencing by double-stranded RNA. Nature 431(7006):343–349. https://doi.org/10.1038/nature02873
23. Morlando M, Ballarino M, Gromak N, Pagano F, Bozzoni I, Proudfoot NJ (2008) Primary microRNA transcripts are processed co-transcriptionally. Nat Struct Mol Biol 15(9):902–909
24. Yin S, Lopez-Gonzalez R, Kunz RC, Gangopadhyay J, Borufka C, Gygi SP, Gao FB, Reed R (2017) Evidence that C9ORF72 dipeptide repeat proteins associate with U2 snRNP to cause Mis-splicing in ALS/FTD patients. Cell Rep 19(11):2244–2256. https://doi.org/10.1016/j.celrep.2017.05.056
25. Dignam JD, Lebovitz RM, Roeder RG (1983) Accurate transcription initiation by RNA polymerase II in a soluble extract from isolated mammalian nuclei. Nucleic Acids Res 11(5):1475–1489. https://doi.org/10.1093/nar/11.5.1475
26. Reddy R, Henning D, Das G, Harless M, Wright D (1987) The capped U6 small nuclear RNA is transcribed by RNA polymerase III. J Biol Chem 262(1):75–81

Chapter 5

Purification of Microprocessor-Associated Factors

Ross A. Cordiner and Sara Macias

Abstract

The Microprocessor complex catalyzes the first step of miRNA biogenesis in the nucleus of mammalian cells. The minimal catalytically active complex is formed by two essential factors, the dsRNA binding protein DGCR8, and the RNase III endonuclease Drosha. Importantly, several co-factors can associate to this complex and modulate the cleavage and binding efficiency of this complex, in a positive or negative manner. Here, we describe a simple method for purification of DGCR8 and Drosha coupled to mass spectrometry or western blot which allows robust identification of unknown associated factors. This approach has recently revealed the presence of a new DGCR8-dependent, Drosha-independent complex involved in RNA turnover.

Key words Immunoprecipitation, Mass spectrometry, Pri-miRNA processing, miRNAs, DGCR8, Drosha, RNA processing, Western blot

1 Introduction

Co-immunoprecipitation or immunoprecipitation (IP) coupled to mass spectrometry techniques aim to identify novel factors associated with the proteins of interest. An important challenge of these approaches is the discrimination amongst true interacting partners versus contaminants. The direct comparison of results obtained with different tagged versions of the same protein, or alternatively, different primary antibodies, plus the inclusion of appropriate negative controls allows the identification of robust interactions and excludes common contaminants obtained by these techniques [1]. Recent efforts towards describing common contaminants in these approaches have been described, such as the "bead proteome" [2] and the "Crapome" [3].

After the discovery of the minimal catalytically active Microprocessor complex [4, 5], many other factors have been shown to modulate the activity as well as the RNA binding capacity of this complex, either by direct protein–protein interactions [6] or

Ulf Andersson Vang Ørom (ed.), *miRNA Biogenesis: Methods and Protocols*, Methods in Molecular Biology, vol. 1823, https://doi.org/10.1007/978-1-4939-8624-8_5, © Springer Science+Business Media, LLC, part of Springer Nature 2018

by binding to the same substrate and precluding or exposing signals necessary for efficient processing [7].

The Microprocessor complex consists of two different essential proteins, the dsRNA binding protein DGCR8 and the RNase III endonuclease Drosha. Here, we present a method to identify new interacting partners for this complex, using either overexpressed or endogenous core Microprocessor components. This approach can be applied to the detailed study of the relative composition of the Microprocessor-associated factors in different cellular contexts, as well as, during different cellular conditions, such as stress responses.

For co-IP of associated factors to overexpressed Microprocessor components, total cell lysates of cells expressing FLAG- or T7-tagged version of DGCR8 and Drosha are prepared and purified using FLAG- or T7-antibody coated beads. After extensive washes of the beads, immunoprecipitates are loaded in a SDS-PAGE gel to be sequenced by mass spectrometry, or analyzed by western blot for specific factors. This protocol is run in parallel with untransfected extracts to discriminate the nonspecific binding of proteins to beads and other reagents used. For the purpose of endogenous proteins purification, total cell extracts are prepared and incubated with beads coupled to primary antibodies against DGCR8 or Drosha, and as a negative control, beads coupled to an IgG antibody from the same serotype are used.

In addition, we have included an RNase treatment step that allows discrimination between those interactions that are RNA-dependent (binding to common RNA substrates) vs independent (direct protein–protein interactions). This approach has been successfully employed to identify many common factors bound to the Microprocessor core, as well as a new DGCR8-dependent, Drosha-independent complex involved in RNA decay [1].

2 Materials and Buffers

2.1 Cell Culture and Transfections

1. Dulbecco's Modified Eagle Medium (DMEM).
2. FCS.
3. Penicillin/Streptomycin.
4. Phosphate buffered saline (PBS) pH 7.4.
5. 0.25% Trypsin-EDTA, Phenol Red.
6. Lipofectamine 2000.
7. Tabletop centrifuge.
8. Refrigerated microfuge.
9. 15 and 50 mL centrifuge tubes.
10. Flag-DGCR8, T7-DGCR8, and Flag-Drosha plasmids.
11. 1.5 mL tubes.

2.2 Cell Lysis and Lysate Clearance

1. Buffer D: 20 mM Hepes-KOH pH 7.9, 100 mM KCl, 0.2 mM EDTA, 5% glycerol.
2. IP buffer: 50 mM Tris pH 7.5, 150 mM NaCl, 1 mM EDTA, 1% Triton ×100.
3. Bioruptor (or probe sonicator).
4. Refrigerated microfuge.
5. 1.5 mL tubes.
6. 1M DTT (1.54 g DTT dissolved in 10 mL of H_2O).
7. PMSF.
8. EDTA-free protease inhibitor cocktail.
9. RQ1 DNase.
10. RNase A.
11. RNase Inhibitors.

2.3 Immunoprecipitation

1. FLAG M2 beads.
2. T7 beads.
3. Protein A magnetic beads.
4. Antibody against DGCR8 (ab90579).
5. Antibody against Drosha (NBP1-03349).
6. Serotype matched IgG control.
7. Rotating wheel.
8. High-salt Buffer D: 20 mM Hepes-KOH pH 7.9, 200 mM KCl, 0.2 mM EDTA, 5% glycerol.
9. IP High-salt buffer: 50 mM Tris pH 7.5, 200 mM NaCl, 1 mM EDTA, 1% Triton ×100.

2.4 SDS-PAGE and Mass Spectrometry/Western Blot

1. Thermomixer or thermoblock.
2. 4× Loading buffer (LD buffer) and 10× Reducing Agent.
3. 4–12% Bis-Tris Novex gels.
4. 20× MOPS running buffer.
5. Prestained Benchmark protein ladder.
6. Gel tank.
7. Western blot transfer unit.
8. ECL and secondary antibodies coupled to HRP.
9. Colloidal blue staining kit.
10. Sterile razor blades.
11. 15 cm sterile plates for gel staining.

2.5 Confirmation of RNase Treatment

1. Thermocycler.
2. Reverse Transcriptase.
3. DNA polymerase.
4. 1× TBE.
5. Agarose.
6. 6× DNA Loading Dye.
7. UV transilluminator.
8. SYBR Safe.

3 Methods

3.1 Cell Culture and Transfections

Detailed below is a simple method to expand and transiently transfect HeLa cells with plasmid vectors expressing Drosha and DGCR8 (FLAG- and/or T7-) and their respective controls using a lipid-based transfection reagent. This protocol has been previously employed to generate enough material to perform immunoprecipitation and mass spectrometry analyses (*see* Fig. 1).

3.1.1 Preparation of Transiently Transfected HeLa Cells Overexpressing DGCR8 and Drosha

1. Seed HeLa cells (or the cell line of choice) into a 6 well plate and expand to 70% confluency. In total, 3 wells of a 6 well plate should be seeded per tagged protein to be transfected (*see* **Note 1**).
2. Briefly rinse cells with PBS, and aspirate the liquid. Add 1.5 mL of DMEM (10% FCS) without antibiotics (*see* **Note 2**), and place cells back in the incubator.
3. Each well of the 6 well plate is to be transfected with 4 μg of plasmid DNA and 10 μL of Lipofectamine 2000 following basic manufacturer's instructions. Cells are to be transfected with the following plasmids: pcDNA3-Flag (empty control plasmid), Flag-Drosha, Flag-DGCR8, pCG-T7 (empty control plasmid), and T7-DGCR8 (*see* **Note 3**).
4. Remove the cells from incubator and add the transfection complex mixture dropwise to each of the 3 wells. The final volume should be 2 mL. Place cells back into the incubator and leave overnight (*see* **Note 4**).
5. After 24 h, trypsinize the cells and pool the transfected cells of the same identity. Reseed the pooled cells in 15 cm plates and allowed to expand for a further 24 h.
6. Collect the cells by trypsinization or scraping 48 h after the initial transfection to allow for optimal protein overexpression. Cells are briefly washed with ice-cold PBS, centrifuged (200 × *g* for 5 min), and cell pellets can be snap-frozen on dry ice and stored at −80 °C until required, or immediately processed.

3.1.2 Preparation of HeLa Whole Cell Extract

1. In total 3 IPs will be performed (IgG control, Drosha, and DGCR8) and 1 × 10 cm dish is at least required per IP. Therefore, expand HeLa cells in 3 × 10 cm tissue culture dishes to 90% confluency.
2. Trypsinize the cells and pool into one centrifuge tube. Cells are then pelleted by centrifugation (5 min at 200 × *g*) in a tabletop centrifuge.
3. Aspirate the supernatant from the pelleted cells and wash the cell pellets in ice-cold 10 mL PBS and spin again at 200 × *g* for 5 min.
4. Aspirate the supernatant and resuspend the cell pellet in 3 mL of ice-cold PBS and transfer 1 mL to 3 × 1.5 mL tube. Pellet cells by spinning at 200 × *g* for 5 min using a microfuge at 4 °C (*see* **Note 5**).
5. Carefully remove supernatant and snap-freeze cell pellet on dry ice and store at −80 °C until required, or process immediately.

3.2 Bead Preparation

Please note that different buffers (IP and Buffer D) and beads (Magnetic and Agarose) are used to purify Microprocessor-associated complexes from transiently transfected cells or HeLa whole cell lysate (*see* Fig. 1).

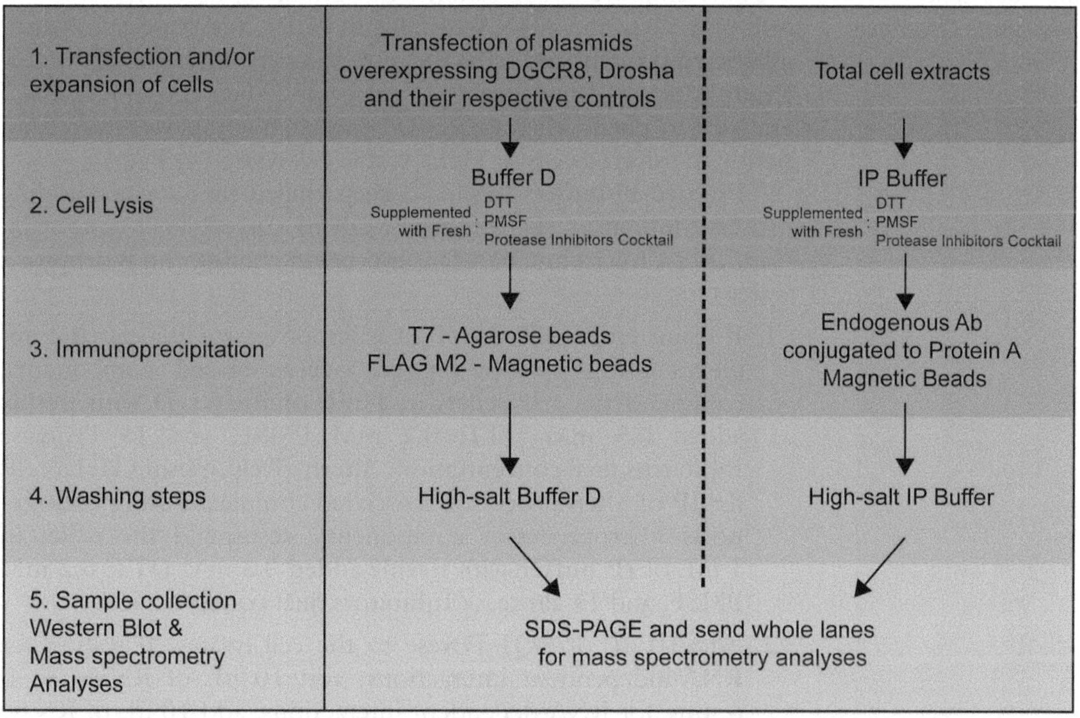

Fig. 1 Experimental overview schematic highlighting different buffers and beads used during the protocol

3.2.1 For Immunoprecipitation of Overexpressed Epitope-Tagged Proteins

1. Pre-wash 10–20 μL of magnetic FLAG or agarose T7 antibody beads per IP by resuspending the beads in 1 mL of buffer D.
2. Microfuge at 1000 × *g* for 30 s, and if using magnetic beads precipitate the beads on a magnetic rack for 30 s.
3. Aspirate as much of the supernatant without disturbing the beads and repeat for a total of three washes (*see* **Note 6**). On last wash resuspend beads in 100 μL of buffer D and keep on ice until required.

3.2.2 For Immunoprecipitation of Endogenous Proteins: Coupling Antibodies to Beads

1. Pre-wash 10–20 μL of Protein A Dynabeads per IP by resuspending the beads in 1 mL of IP buffer.
2. Microfuge at 1000 × *g* for 10 s and precipitate the beads on a magnetic rack.
3. After aspirating the supernatant (*see* **Note 6**), repeat for a total of three washes. On last wash resuspend beads in 500 μL of IP buffer and add 1 μg of the primary antibody of your choice and include the negative species matched IgG as a control. Incubate for 3–4 h at 4 °C with end-over-end rotation.
4. Microfuge at 1000 × *g* for 30 s and magnetically precipitate the beads. Perform three washes with 1 mL of IP buffer, and resuspend the beads in 100 μL of IP buffer and keep on ice until needed.

3.3 Cell Lysis and Lysate Clearance

We have employed a sonication-based cell disruption method to lyse cells and shear their DNA content in order to release chromatin-associated proteins and reduce sample viscosity. Please note that different buffers (IP and Buffer D) and beads (Magnetic and Agarose) are used to purify Microprocessor-associated complexes from transiently transfected cells or HeLa whole cell lysate (*see* Fig. 1).

Prepare all buffers fresh and keep chilled on ice when adding protease inhibitors, DTT, and PMSF to correct final concentration, cool a microfuge to 4 °C, and begin chilling the Bioruptor's water tank.

1. If using HeLa cells for IP of Microprocessor-associated complexes using **epitope-tagged overexpressed** components, resuspend the cell pellets in 1 mL of **Buffer D** with freshly added 0.5 mM DTT, 0.2 mM PMSF, and 1× Protease Inhibitors final concentration. Alternatively, if using HeLa cells for IP of Microprocessor-associated complexes using **endogenous** Microprocessor components, resuspend the pellet in 1 mL of **IP buffer** with freshly added 0.5 mM DTT, 0.2 mM PMSF, and 1× Protease Inhibitors final concentration.
2. Add 10 μL of RQ1 DNase to the cell lysates. If testing for RNA-independent interactions, add 10 μL of RNase A. If testing for RNA-dependent interactions, add 10 μL of RNase Inhibitor.

3. Sonicate in the Bioruptor (Diagenode) using the low setting for a total of five cycles of 30 s on/30 s off (*see* **Note 7**).
4. Spin at 16,000 × *g* for 10 min using a microfuge at 4 °C, and transfer the supernatant to a new 1.5 mL tube.
5. Microfuge samples again at 16,000 × *g* for 10 min at 4 °C, and transfer 95% of the supernatant to a new 1.5 mL tube, taking care not to disturb any pelleted debris and store on ice (*see* **Note 8**).
6. Transfer 50 μL (5%) of the cleared cell lysate to 1.5 mL tube labeled INPUT and freeze on dry ice. This can be processed later for input analyses.

3.4 Immunoprecipitation and Washes

1. Microfuge the beads at 1000 × *g* for 30 s to initially pellet the beads (*see* from Subheadings 3.2.1 and 3.2.2), and then magnetically precipitate the magnetic beads. Fully aspirate the last wash, being careful not to disturb beads.
2. Transfer the cleared cell lysate to the 1.5 mL tube containing the beads.
3. Rotate the immunoprecipitations end-over-end at 4 °C from 1 h to overnight (*see* **Note 9**). The overexpressed protein IPs can be carried out at shorter times, whereas endogenous IPs are usually incubated overnight.
4. Microfuge the IPs at 1000 × *g* for 30 s to collect beads at the base of the tube (*see* **Note 10**), and then magnetically precipitate the magnetic beads.
5. Remove 50 μL and transfer to a 1.5 mL tube labeled UNBOUND and immediately snap-freeze on dry ice (*see* **Note 11**).
6. Aspirate remaining lysate from IPs being careful not to disturb beads.

At this step, the standard salt buffers are replaced with high-salt IP buffer (200 mM NaCl) or high-salt buffer D (200 mM KCl) to perform high-stringency washes with rotation for 5 min. Please note that different buffers are used for different IPs (*see* Fig. 1).

7. To wash beads, resuspend in 1 mL of high-salt buffer, and rotate at room temperature for 5 min at 16 rpm. Microfuge for 10 s at 1000 × *g* to pellet beads, and if using magnetic beads precipitate them on a magnetic rack. Fully aspirate the wash buffer.
8. Repeat **step 7** for a total of five washes. After rotating for the last wash, remove 100 μL of bead suspension and place in a clean 1.5 mL tube labeled 10% IP. This can be used to assess IP efficiency.

9. Microfuge all tubes for 10 s at 1000 × *g* and then magnetically precipitate the magnetic beads. Completely aspirate the last wash buffer and resuspend the beads in 15 μL of 2× LD buffer with 2× Reducing agent (7.5 μL 4× LD buffer, 3 μL 10× Reducing agent, and 4.5 μL of PBS) and incubate at 70 °C for 10 min in a thermomixer with 1100 rpm shaking.
10. If IPs are to be stored and not processed immediately, microfuge for 10 s at 1000 × *g* to pellet beads, and then magnetically precipitate the beads from the sample and transfer the eluate to a labeled 1.5 mL tube and snap-freeze on dry ice. This can then be stored at −80 °C until required (*see* **Note 12**).

3.5 SDS PAGE and Sample Isolation

To obtain the complete interactome of Microprocessor-associated components, we chose to sequence whole lanes from the IPs. For this, it is necessary to perform the staining and the gel excision steps within a laminar hood to prevent contamination from aerosols.

1. Spin the 1.5 mL tubes at 1000 × *g* for 10 s to collect liquid at the base of the tube and heat at 70 °C for 10 min. Load a prestained protein ladder, and all of the 15 μL mass spectrometry IPs into a 10 well 1.0 mm 4–12% Bis-Tris Novex gel (*see* **Note 13**). Leave at least one empty lane between samples to prevent cross contamination when cutting lanes.
2. Run the gel at 170 V for 5 min or until the samples have entered the gel and begin to separate. This can be judged by separation of the protein marker and when the dye front reaches the first graduation of the gel case (*see* Fig. 2a). For confirmation of optimal immunoprecipitation conditions, 10% of the IP is run in a separate gel and stained with colloidal coomassie blue prior to send samples to mass spectrometry analyses (*see* Fig. 2b).
3. Transfer the unopened Novex gel cassette to a flow laminar hood. Break open the cassette and transfer gel to a clean 15 cm dish and stain with colloidal coomassie blue staining kit as per manufacturer's instructions.
4. After visualizing the lanes with protein stain, return the gel to the laminar hood and use a clean blade to excise the whole stained lane from the gel (as seen in Fig. 2a), and transfer to a labeled sterile tube and store at 4 °C. For each individual sample, use a new pair of gloves and razor blade to prevent cross contamination.
5. Send the individual mixture of proteins for mass spectrometry analyses.

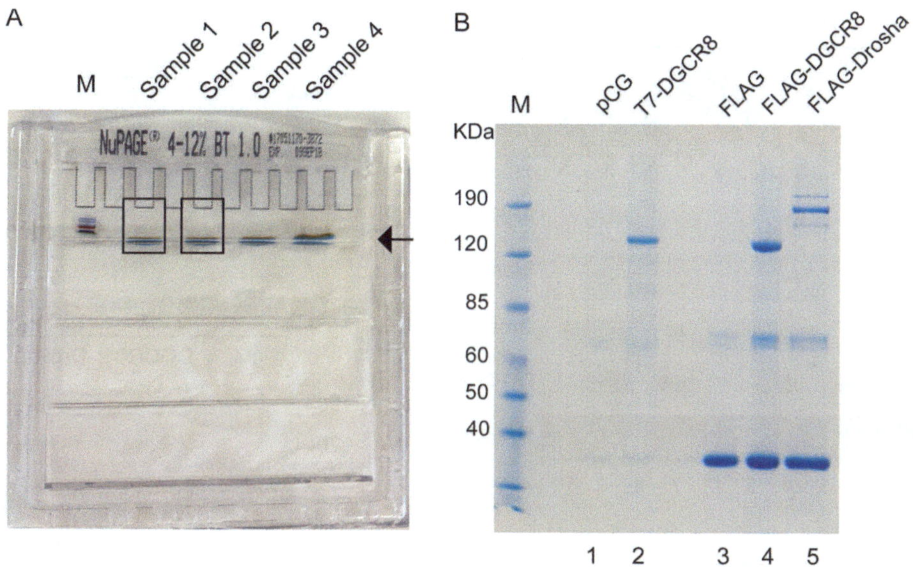

Fig. 2 Purification of associated factors with overexpressed Microprocessor components DGCR8 and Drosha. (**a**) SDS-PAGE separation of immunoprecipitates after 5 min running at 170V or until the protein dye front reaches the first graduation on the Novex gel cassette. Samples are sliced as depicted by the black rectangle. (**b**) Coomassie blue staining of 10% of the immunopurified complexes with overexpressed T7-DGCR8, FLAG-DGCR8, and FLAG-Drosha (lanes 2, 4, and 5). As negative controls, cells transfected with control plasmids were immunoprecipitated using T7 (pCG-T7, lane 1) and FLAG antibody coated beads (pcDNA3-FLAG, lane 3). M denotes marker (reproduced from (1) under a CC BY 4.0, DOI 10.1016/j.molcel.2015.11.011)

3.6 Input Processing and Analysis Controls

3.6.1 Analysis of Bona fide Interactions by Western Blot

To define whether the IPs have good signal to noise ratios and are robust enough, staining and western blot analysis on samples of the cell lysate before and after IP to check for antigen depletion are performed. Co-IP of bona fide protein interactors by western blot, such as reciprocal DGCR8 and Drosha co-immunoprecipitation is also used as a control for good experimental methodology. In addition, confirmation of RNase treatment can be assessed by analyzing RNA content from the UNBOUND fraction (*see* Fig. 3).

1. Defrost INPUT, UNBOUND, and 10% IP samples on ice.
2. Transfer 15 μL of the lysates (INPUT and UNBOUND) to a new 1.5 mL tube and add 7.25 μL of 4× LD dye and 3 μL of 10× Reducing agent, mix, and centrifuge briefly to collect liquid at the base of the tube. Do the same for the 10% IP sample and incubate all samples for 10 min at 70 °C with 1100 rpm agitation.
3. Briefly centrifuge tubes after heating and place on a magnetic stand to precipitate the beads. Load the supernatant into a 12 well 4–12% Bis-Tris Novex gel, along with 10 μL of protein ladder. The gel should be run in 1× MOPS for 1 h at 150 V, or until the dye front has reached the bottom of the gel.

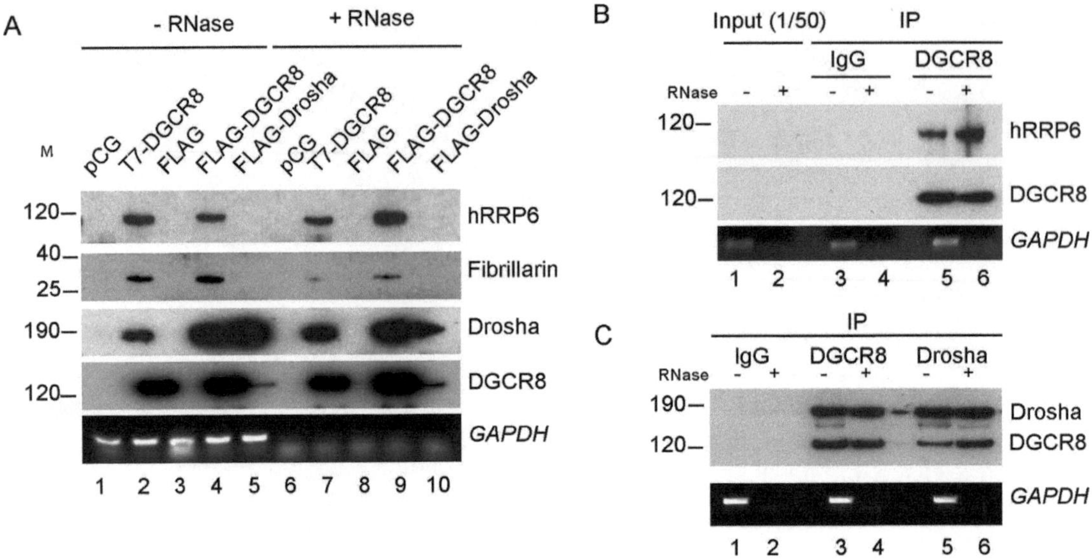

Fig. 3 Immunoprecipitation followed by western blot analyses with overexpressed and endogenous Microprocessor factors. (**a**) Validation of proteins interacting with T7-DGCR8, FLAG-DGCR8, and FLAG-Drosha by immunoprecipitation followed by western blot analysis with specific antibodies, in the presence (lanes 6–10) or absence of RNase A (lanes 1–5). The exosome component hRRP6 and Drosha are co-immunoprecipitated with DGCR8 independent of the presence of RNA, whereas fibrillarin co-immunoprecipitation depends on RNA presence. The RT-PCR amplification of *GAPDH* serves as a control for RNase treatment (bottom panel), (**b, c**) The co-immunoprecipitation of hRRP6 (**b**, compare lane 5 and 6) and Drosha (**c**, compare lanes 3 and 4) with DGCR8 is independent of the presence of RNA. Drosha co-immunoprecipitates DGCR8 independently of the presence of RNA (**c**, lanes 5 and 6). *GAPDH* mRNA PCR amplification serves as a control for RNase treatment. (a and b are reproduced from (1) under a CC BY 4.0, DOI 10.1016/j.molcel.2015.11.011)

4. If assessing IP samples for noise, stain the gel using colloidal blue staining kit as per the manufacturer's instructions (*see* Fig. 2b). For assessment of co-IP of bona fide Microprocessor-associated proteins by western blot, transfer the gel to a nitrocellulose membrane using a blotting system (Fig. 3a, for overexpressed proteins, and Fig. 3b, c for endogenous proteins).
5. Block the membrane by agitating it in 5% non-fat powder PBS-T (PBS supplemented with Tween-20 to a final concentration of 0.05%) for 30 min to 1 h at room temperature.
6. Dilute the appropriate primary antibody (usually 1:500–1:1000 for anti-Drosha and DGCR8, and 1:10,000 for overexpressed proteins for anti-FLAG or T7) in 5% non-fat powder PBS-T and leave to incubate overnight, rotating at 4 °C.
7. Perform 3 PBS-T washes with agitation for 10 min each, at room temperature.
8. Dilute the appropriate secondary antibody in 5% non-fat powder PBS-T and leave to incubate at room temperature with agitation for over 1 h.

9. Perform 3 PBS-T washes with agitation for 20 min each, at room temperature.
10. Image western blot with ECL and chemiluminescent film (*see* Fig. 3).

3.6.2 Confirmation of RNase Treatment

1. Take the UNBOUND fraction (50 μL) and perform a Trizol LS extraction, or alternatively Proteinase K treatment followed by an acidic phenol-chloroform extraction. After precipitation with ethanol or isopropanol, resuspend pellets in 10 μL of RNase-free water.
2. Perform retrotranscription against an abundant house-keeping gene, such as *GAPDH*, *RN7SK*, or *ACTIN*. Amplify the transcript of interest by PCR. Next, run the PCR products in a 1% TBE-agarose gel and stain with SYBR safe to check for efficient RNase treatment (as seen in Fig. 3a–c bottom panels).

4 Notes

1. To achieve 70% confluency 20–24 h after seeding, seed 0.18–0.2×10^6 cells per well of a 6 well plate. This equates to roughly 1/48th of a confluent T75 flask of cells (250 μL of 12 mL suspension).
2. Using antibiotics in combination with lipid-based transfection reagents can be highly cytotoxic and negatively affect transfection efficiency of your experiment.
3. Ensure plasmid DNA concentration is at least 0.5 μg/μL. Adding large volumes low concentration plasmid DNA can carry over contaminants from the plasmid DNA prep, which can negatively affect transfection efficiency.
4. Transfection complex can be removed from cells after 6 h. This may increase cell viability and has no effect on transfection efficiency.
5. Doing this ensures that equal volumes of protein are used for the individual IP and avoids the need to normalize the input material.
6. Accidently aspirating the magnetic beads is not usually a problem as they are isolated against the tube wall away from the base of the tube. However, agarose beads will remain pelleted at the bottom. To aid aspiration we recommend placing a p1000 tip into a p10 tip to improve aspiration accuracy and reduce turbulent flow of supernatant into the pipette, which could result in bead loss.
7. If snotty DNA pellet remains in the lysate after sonication, this can be due to the cell pellet being too concentrated. Resuspend

in larger volume, or alternatively use a Branson probe sonicator (five cycles of 10 s on followed by 30 s off using 10 AMP) to lyse cells and shear DNA. It is important to shear the cellular DNA content as this will aid the release of chromatin-associated proteins.

8. Proteus columns can be used for further cell lysate clearance if required.
9. The incubation time depends on quality of the antibody used and/or the abundance of the antigen in the cell lysate. Adjust incubation times depending on these factors. To assess IP quality, *see* **Note 11**.
10. Pelleting beads before applying to a magnet aids magnetic precipitation and clears liquid/beads from the tube cap.
11. Western blot analyses can be used to assess IP efficiency by comparing protein depletion of equivalent fractions of the UNBOUND with the INPUT samples. The UNBOUND also serves to monitor optimal RNase treatment.
12. The use of buffer D, which is high in glycerol, allows freezing of "active" immunoprecipitates with overexpressed proteins, which can be used in "in vitro" processing assays.
13. Do not use homemade SDS-PAGE running buffer here, rather segregate a commercially available premade buffer specifically for your mass spectrometry analyses to prevent contamination.

References

1. Macias S, Cordiner RA, Gautier P, Plass M, Caceres JF (2015) DGCR8 acts as an adaptor for the exosome complex to degrade double-stranded structured RNAs. Mol Cell 60:873–885
2. Trinkle-Mulcahy L, Boulon S, Lam YW, Urcia R, Boisvert FM, Vandermoere F, Morrice NA, Swift S, Rothbauer U, Leonhardt H et al (2008) Identifying specific protein interaction partners using quantitative mass spectrometry and bead proteomes. J Cell Biol 183:223–239
3. Mellacheruvu D, Wright Z, Couzens AL, Lambert J-P, St-Denis NA, Li T, Miteva YV, Hauri S, Sardiu ME, Low TY et al (2013) The CRAPome: a contaminant repository for affinity purification–mass spectrometry data. Nat Methods 10:730–736
4. Gregory RI, Yan K-P, Amuthan G, Chendrimada T, Doratotaj B, Cooch N, Shiekhattar R (2004) The microprocessor complex mediates the genesis of microRNAs. Nature 432:235–240
5. Landthaler M, Yalcin A, Tuschl T (2004) The human DiGeorge syndrome critical region gene 8 and its D. melanogaster homolog are required for miRNA biogenesis. Curr Biol 14:2162–2167
6. Cheng TL, Wang Z, Liao Q, Zhu Y, Zhou WH, Xu W, Qiu Z (2014) MeCP2 suppresses nuclear MicroRNA processing and dendritic growth by regulating the DGCR8/drosha complex. Dev Cell 28:547–560
7. Fernandez N, Cordiner RA, Young RS, Hug N, Macias S, Cáceres JF (2017) Genetic variation and RNA structure regulate microRNA biogenesis. Nat Commun 8:15114

Chapter 6

Inhibiting Pri-miRNA Processing with Target Site Blockers

Annita Louloupi and Ulf Andersson Vang Ørom

Abstract

The pri-miRNA processing is important for the final regulatory role of miRNAs on the expression of their target transcripts. The processing variability between pri-miRNAs can determine the final miRNA abundance better than primary transcription itself. Thus studying the in vivo pri-miRNA biogenesis could give more insights into the contribution of each individual miRNA on regulation of gene expression. Interfering processing of a specific pri-miRNA has been challenging due to the nature of the current RNA interfence methods. Here, we describe step by step a method to arrest processing of specific pri-miRNAs in vivo using LNA microRNA Target Site Blockers. We explain in detail the various aspects of this approach that can easily be applied to different mammalian cell types. The nature of this protocol allows the in vivo study of pri-miRNA processing and processing kinetics in cells treated with different conditions, mutants, and/or cancer cell lines under physiological conditions.

Key words Pri-miRNA processing dynamics, BrU pulse-chase labeling

1 Introduction

miRNA biogenesis starts with the transcription of the primary miRNA (pri-miRNA) by RNA polymerase II [1]. Due to sequence composition the pri-miRNA forms a stem-loop structure that is recognized by the microprocessor complex, comprising Drosha and DGCR8. More specifically, the microprocessor recognizes the single-stranded RNA tails, the stem of ~35 bp in length, and a terminal loop of the pri-miRNA [2]. Microprocessor acts as ruler to measure ~11 bp from the basal junction and ~22 bp from the apical junction where Drosha performs the cleavage. Then Dicer recognizes pre-miRNA where the sequence ends of pre-miRNA are recognized by the PAZ (PIWI–Argonaute–ZWILLE) domain of human Dicer, containing two basic domains: one that interacts with the 5′-phosphorylated end of the pre-miRNA and one that interacts with the 3′ end. Dicer generates two ~22 nt strands that pair to each other with ~2 nt 3′ overhangs [3]. The mature microRNA is then one of the strands that according to its thermodynamics would be loaded into an Argonaute protein to

Ulf Andersson Vang Ørom (ed.), *miRNA Biogenesis: Methods and Protocols*, Methods in Molecular Biology, vol. 1823, https://doi.org/10.1007/978-1-4939-8624-8_6,

form the core of the silencing complex and the other strand is discarded [4].

The sequence composition of pri-miRNA is important for the efficiency of the pri-miRNA processing [5]. More specifically, the positions −14 and −13 (+1 first nt of mature miRNA) are typically occupied by a U and G, respectively. The CNNC motif (N signifies any nucleotide) found at 17–18 nt downstream of Droshas cleavage site contributes to pri-miRNA recognition and enhances pri–miRNA processing. More sequence features for different pri-miRNAs can contribute to the processing and with the method described here we could identify them.

The idea is based on the application of commercial miRCURY LNA™ microRNA Target Site Blockers (TSB), which are typically used to compete with RISC complex for a microRNA target site of a particular mRNA, to attenuate the pri-miRNA processing. These antisense oligonucleotides can be transfected using commercial transfection reagents into cells. While entering the nucleus, the TSB will recognize and strongly hybridize complementary sequences preventing the microprocessor complex or other proteins from interacting with the pri-miRNA. In addition, the biochemistry of the TSB ensures that the pri–miRNA target sequence will not be degraded by RNase H-dependent degradation (Fig. 1).

The protocol described above can be used to answer questions regarding pri-miRNA processing and to identify more pri-miRNA sequence features important for the miRNA expression. We have successfully used this approach to attenuate the maturation of miRNAs and we highly recommend it since it is a relatively simple method that can be used in different cell types under different biological conditions.

2 Materials

2.1 Lab Equipment

1. Cell culture 6-well plates.
2. Laminar flow hood.
3. Cell culture incubator.
4. Refrigerated microcentrifuge.
5. 1.5 mL Eppendorf tubes.

2.2 Chemicals

1. DMEM, high glucose, no glutamine, no phenol red.
2. Fetal Bovine Serum.
3. Opti-MEM I Reduced Serum Medium.
4. PBS, pH 7.4.
5. TRIzol Reagent.
6. GlycoBlue.
7. Chloroform.

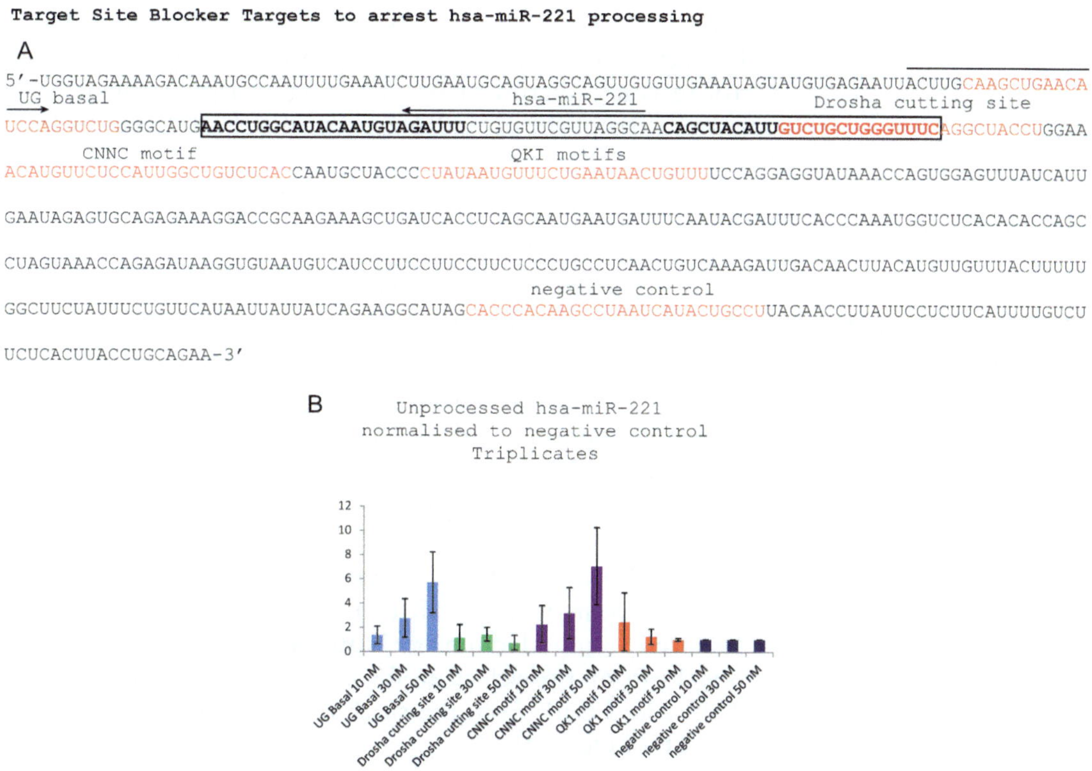

Fig. 1 (**A**) Target sequences of the miRCURY LNA blockers (red); arrows indicate the primers that were used to detect the unprocessed hsa-miR-221 (**B**) qPCR results of unprocessed hsamiR-221 normalized to negative control sites in triplicates

8. Isoamyl alcohol.
9. Isopropanol.
10. Ethanol.
11. 3 M sodium acetate (pH 5.2).
12. Distilled water.
13. miRCURY LNA Target Site Blockers from Exiqon.

2.3 Kits

1. Reverse Transcriptase.
2. qPCR Master mix.
3. Transfection Reagent.

3 Methods

3.1 Design of Pri-miRNA Processing Blockers

TSB were designed by providing 23–26 nucleotides long target sequences to Exiqon. The target sequence should include important sequence residues for the pri-miRNA processing such as the 5p Drosha cleavage site (the 5′ end of the miRNA in the pre-miRNA) and the 3p Drosha cleavage site (the 3′ end of the miRNA in the pre-miRNA). Additional residues such as the UG basal sequence (~ −13, −14) and CNNC sequence (+17) could also be used as target sequence for the TSB. Always include a negative control in your experiment by providing a target sequence that does not interrupt the pri-miRNA processing (>200 nt away from the pri-miRNA sequence).

Upon arrival resuspend the TSB to a final concentration of 50 μM.

3.2 Cell Culture Transfection of Target Site Blockers

1. Grow HEK293 cells in a DMEM growth medium with 10% Fetal Bovine Serum (FBS) under normal growth conditions (typically 37°C and 5% CO_2) (*see* **Note 1**).
2. The day before TSB transfection, seed 0.5 × 10^6 cells per well in 6-well plate in 2 mL cell culture media DMEM supplemented with 10% FBS. Shake the plate gently to ensure cells will not form clamps in the center of the well. Seed enough wells for negative controls (*see* **Notes 2** and **3**).
3. On the day of transfection, prepare a master mix (for triplicates) of your desired TSB concentration in Opti-MEM™ I Reduced Serum Medium (*see* **Notes 4** and **5**). For example for 3 wells from a 6-well plate for final concentration of 50 nM TSB, add 3.5 μL from the stock (50 μM) in (3.5 × 100 μL) 350 μL of Opti-MEM™.
4. Mix well by vortexing.
5. Add HiPerfect Transfection Reagent (*see* **Note 6**) directly to the master mix (3.5 × 6 μL).
6. Vortex the master mix and incubate at room temperature for 15 min.
7. In the meanwhile, replace the old media (2 mL per well) with 900 μL DMEM supplemented with 10% FBS per well.
8. Add 100 μL from the master mix to each well dropwise and gently rock the plate.
9. Incubate the plate for 6–15 h in the cell culture incubator under normal growth conditions (37°C and 5% CO_2).
10. Add 1 mL of DMEM supplemented with 10% FBS in each well and continue incubating the cells for a total of 24 h (after transfection).
11. Remove the media and wash cells once with 1× PBS (*see* **Note 7**).
12. Continue with RNA isolation by adding 1 mL of Trizol directly to the cells.

3.3 Isolation of RNA

1. Incubate cells in Trizol for 1–2 min at room temperature to allow the cell lysis to occur. Pipet up and down to lyse the cells completely and transfer the lysate in an 1.5 mL Eppendorf tube.

2. In each tube add 0.2 mL of 24:1 chloroform:isoamyl alcohol (0.2 mL per 1 mL Trizol) and vortex for 10 s.
3. Centrifuge tubes at 4°C for 20 min at full speed.
4. Transfer the upper aqueous layer to a new tube and perform another round of 24:1 chloroform:isoamyl alcohol extraction to remove any traces of phenol (*see* **Note 8**).
5. Transfer the upper aqueous layer to a new tube and add 1 mL isopropanol (0.5 mL per 1 mL Trizol).
6. Mix the solution well and allow RNA to precipitate for 0.5–1 h at −20°C.
7. Collect the precipitated RNA by centrifugation at full speed for 30 min at 4°C.
8. Remove supernatant and wash the pellet twice with 75% ethanol, centrifuge again, and remove any remaining ethanol.
9. Resuspend the pellet in 80 μL RNase-free water.
10. Store samples at −80°C unless immediately proceeding with cDNA preparation.

3.4 Primer Design for RT-qPCR to Detect the Unprocessed and Processed Pri-miRNA Transcript and Transcript Quantification by Reverse Transcription Followed by Quantitative PCR

1. Detect the mature miRNA and design one of your primers to anneal at the cleavage site or design the primer to produce an amplicon surrounding Droshas cleavage site (*see* **Note 9**).
2. Prepare RT reactions using GoScript Reverse Transcriptase (cat.no A5003 Promega). Use 100–1000 ng as input for reverse transcription and perform the reaction according to manufacturer's instructions. The cDNA was quantified on a 7900HT Fast Real-Time PCR system (Applied Biosystems) using the GoTaq qPCR Master mix (cat.no A6001 Promega) (*see* **Notes 10** and **11**).

4 Notes

1. Different cell lines can be used as well.
2. We advise to use an antibiotic-free medium to avoid any potential toxic effect caused during transfection.
3. The indicated number of cells can be adjusted for different cell lines according to size and rate of growth. Cells should be 60–70% confluent on the day of transfection.
4. All the experiments should be performed under a tissue culture hood. TSB oligonucleotides are susceptible to degradation by exogenous nucleases introduced during handling. Wear powder-free gloves when handling this product and use DNase-free reagents and filter pipette tips.

5. In order to find the best TSB final concentration for the transfection, we advise to perform a pilot experiment to identify the suitable amount of TSB for each pri-miRNA. On our hands dilutions starting from 10 to 50 nM had the best result without off-target effects.
6. Vortex well before use the HiPerfect Transfection Reagent. On our hands 6 μL of the reagent is enough for a well in 6-well plate with HEK293 cells. We advise to calibrate the amount of the HiPerfect Transfection Reagent for different cell lines and number of cells and media volume.
7. In this step, PBS can be added to the cells and transfer the plates to a fume hood where Trizol can be added.
8. To obtain very pure RNA from Trizol samples in this step, the upper aqueous layer can be extracted a second time using acidic phenol pH 4.5–5 1:1 volume followed by another extraction with 24:1 chloroform:isoamyl alcohol to remove any traces of phenol. The DNA remains in the organic phase in the acidic phenol extraction.
9. The 5′ end of 5p miRNA and 3′ end of 3p miRNA are generally assumed as DROSHA cleavage site [2].
10. Any cDNA and RT-qPCR kit can be used according to your preference.
11. The starting RNA amount for the cDNA synthesis can be optimized according to the expression of your pri-miRNA.

References

1. Lee Y, Kim M, Han J, Yeom KH, Lee S, Baek SH, Kim VN (2004) MicroRNA genes are transcribed by RNA polymerase II. EMBO J 23(20):4051–4060. https://doi.org/10.1038/sj.emboj.7600385
2. Han J, Lee Y, Yeom KH, Nam JW, Heo I, Rhee JK, Sohn SY, Cho Y, Zhang BT, Kim VN (2006) Molecular basis for the recognition of primary microRNAs by the Drosha-DGCR8 complex. Cell 125(5):887–901. https://doi.org/10.1016/j.cell.2006.03.043
3. Lee Y, Ahn C, Han J, Choi H, Kim J, Yim J, Lee J, Provost P, Radmark O, Kim S, Kim VN (2003) The nuclear RNase III Drosha initiates microRNA processing. Nature 425(6956):415–419
4. Schwarz DS, Hutvagner G, Du T, Xu Z, Aronin N, Zamore PD (2003) Asymmetry in the assembly of the RNAi enzyme complex. Cell 115(2):199–208
5. Auyeung VC, Ulitsky I, McGeary SE, Bartel DP (2013) Beyond secondary structure: primary-sequence determinants license pri-miRNA hairpins for processing. Cell 152(4):844–858. https://doi.org/10.1016/j.cell.2013.01.031

Chapter 7

MicroRNA Analysis Using the Quantitative Real-Time PCR Reaction

Marta Kotlarek, Anna Kubiak, Krystian Jażdżewski, and Anna Wójcicka

Abstract

The analysis of microRNA expression patterns provides new insights into numerous cellular processes and their aberrances in diseases. Despite its potential pitfalls, the quantitative real-time polymerase chain reaction (qPCR) is the most commonly used tool for microRNA profiling. The method requires extraction and quality analysis of RNA, which is further reverse transcribed using specific primers and used as a template in a qPCR reaction. All these elements have been addressed in this chapter.

Key words microRNA identification, microRNA quantification, TaqMan probe, qPCR, Stem-loop primer, RNA extraction, FFPE samples, Serum samples

1 Introduction

The quantitative real-time polymerase chain reaction (qPCR) is the most commonly used tool for microRNA profiling, thanks to its relative ease of use, high availability and cost-effectiveness. The method has been long considered a gold standard for RNA quantification [1]. However, the analysis of miRNAs is complicated by the short length and high sequence similarity of the molecules, requiring the design of specific stem-loop primers for reverse transcription and probes for further quantification of cDNA [2]. Numerous companies provide ready-to-use assays for miRNA quantification, but the specific primers and probes can also be custom-designed, what allows for the analysis of newly discovered microRNAs and microRNA isoforms [3]. MicroRNA quantification in qPCR is possible with even small amounts of RNA for analysis, what is especially useful when sample is very limited. However, the method has serious limitations with respect to discrimination between microRNA length and sequence isoforms [4–7] what can affect the obtained results.

Ulf Andersson Vang Ørom (ed.), *miRNA Biogenesis: Methods and Protocols*, Methods in Molecular Biology, vol. 1823, https://doi.org/10.1007/978-1-4939-8624-8_7,

2 Materials

Prepare all solutions using ultrapure water (prepared by purifying deionized water) and analytical grade reagents. Prepare and store all reagents at room temperature, unless otherwise indicated in the text.

2.1 Laboratory Facilities and Equipment

1. Set of pipettes with adjustable dispensing volume:
   (a) 0.1–2.5 μL.
   (b) 2–20 μL.
   (c) 20–200 μL.
   (d) 100–1000 μL.
2. Filter tips for each volume pipette.
3. Thermal cycler.
4. Spectrophotometer.
5. Thermoblock/water bath with a temperature range up to 85°C.
6. Real-time PCR instrument.
7. Centrifuge up to 20,000 × *g*.
8. Tissue homogenizer.
9. Mini spin centrifuge.
10. High quality 1.5 mL and 2 mL tubes and PCR tubes.
11. Adhesive sealing foil for thermal cycler.
12. Hood.
13. Laboratory ice machine.

2.2 General Reagents for Molecular Biology

1. Xylene.
2. 100–96% ethanol.
3. Glycol.
4. 10 M NaOH.
5. 200 mM Tris-HCl, pH 7.
6. 20 mg/μL Proteinase K.
7. 0.1 M Dithiothreitol (DTT).
8. 3 M Sodium acetate.

2.3 Kits and Reagents for miRNA Extraction and Analysis

1. RNA extraction from fresh/frozen tissue samples and cell cultures: TRIzol Reagent.
2. RNA extraction from formalin-fixed paraffin-embedded tissues and cytological slides: InviTrap Spin Universal RNA Mini Kit.

3. RNA extraction from plasma: mirVana miRNA Isolation Kit, with phenol.
4. MicroRNA Reverse Transcription Kit.
5. TaqMan MicroRNA Assay (RT primer).
6. Custom-designed RT-PCR primers and probes.
7. Real-time PCR system.
8. Multiwell Plate 96 or 386, white.
9. Sealing Foil.
10. TaqMan MicroRNA Assay (probe and primer).
11. TaqMan 2× Universal PCR Master Mix.

3 Methods

Perform all the procedures at room temperature, unless otherwise indicated in the text.

3.1 RNA Extraction from Fresh/Frozen Tissue or Cell Cultures Using the TRIzol Reagent

1. In a single Eppendorf tube, add 750 μL of TRIzol to 50–100 mg of tissue or to 10^5–10^7 monolayer cells and homogenize.
2. Incubate for 5 min, then add 400 μL of chloroform.
3. Mix by inverting tubes 2–3 times, incubate for 5 min.
4. Centrifuge for 15 min at 12,000 × *g* at 4 °C.
5. Transfer the upper aqueous phase containing RNA to a new tube and add 400 μL of chloroform.
6. Mix by inverting the tubes 2–3 times, incubate for 5 min.
7. Centrifuge for 15 min at 12,000 × *g* at 4 °C.
8. Transfer the upper aqueous phase to a new tube and add 600 μL of 96–100% ethanol with 0.3 M sodium acetate. Optionally store at −20 °C overnight.
9. Centrifuge for 10 min at 12,000 × *g* at 4 °C.
10. Discard the supernatant.
11. Wash the pellet in 1 mL of 75% ethanol and centrifuge for 5 min at 7500 × *g* at 4 °C.
12. Discard the supernatant and repeat wash.
13. Discard the supernatant, leave RNA on ice until dry.
14. Suspend RNA in nuclease-free water.
15. Keep RNA on ice or store at −80 °C.

3.2 RNA Extraction from Formalin-Fixed Paraffin-Embedded Tissue Samples Using the InviTrap Spin Universal RNA Mini Kit

1. Preheat the thermoblock/water bath to 48 °C.
2. Cut the FFPE block into 6–10 sections up to 10 μm and transfer to a 2 mL tube.
3. Add 0.5 mL of xylene. Mix until paraffin is dissolved.
4. Centrifuge for 2 min at a maximum speed.
5. Discard the supernatant and wash the pellet in 1 mL of 96–100% ethanol.
6. Centrifuge shortly at a maximum speed, discard ethanol and leave the sample at 48 °C to evaporate.
7. Add 180 μL of RNase-free water, 40 μL of proteinase K (20 mg/mL), and 2.2 μL of 0.1 M DTT, and homogenize.
8. Incubate for 10 min at 48 °C and then heat to 80 °C for 10 min.
9. Add 900 μL of Lysis Solution TR (containing 0.01 M DTT) and ten Zirconia Beads II, vortex for 5 min. Transfer the supernatant to a new 2 mL tube. Leave overnight at −20 °C.
10. Thaw the samples on ice, centrifuge for 2 min at maximum speed. Transfer the supernatant to a new 2 mL tube, then add 0.7× volume of 100% ethanol. Gently pipette.
11. Transfer the lysate onto the RTA spin filter (maximum 800 μL).
12. Centrifuge for 2 min at 11,000 × *g*.
13. Discard the flow-through and replace with a new receiver tube.
14. Transfer the last part of the lysate onto the RTA spin filter and centrifuge as described above.
15. Wash the spin filter in 600 μL of Wash Buffer R1, centrifuge at 11,000 × *g* for 1 min, discard the flow-through.
16. Wash the spin filter in 700 μL of Wash Buffer R2, centrifuge at 11,000 × *g* for 1 min, discard the flow-through, repeat the washing step.
17. Centrifuge the spin filter for 4 min at a maximum speed to eliminate any residual ethanol.
18. Discard the RTA Receiver tube. Transfer spin filter set to a new 1.5 mL tube and add 40 μL of nuclease-free water.
19. Wait for 2 min and centrifuge at 11,000 × *g* for 1 min.
20. Keep RNA on ice or store at −80 °C.

3.3 RNA Extraction from Cytological Slides Using the InviTrap Spin Universal RNA Mini Kit

1. Soak the slide in xylene and mix under the fume hood until the cover glass is removed (from 2 h to overnight).
2. Preheat the thermoblock/water bath to 52 °C.
3. Remove the slide from xylene and scrape the tissue material off with a scalpel to a 1.5 mL tube.
4. Wash the pellet with 500 μL of 96% ethanol.
5. Centrifuge shortly, discard ethanol and leave the sample at 52 °C to evaporate.
6. Add 180 μL of RNase-free water, 40 μL of proteinase K (20 mg/mL), and 2.2 μL of 0.1 M DTT.
7. Incubate for 10 min at 48 °C and then for 10 min at 80 °C.
8. Add 900 μL of Lysis Solution TR (containing 0.01 M DTT), vortex for 5 min. Transfer the supernatant to a new 2 mL tube. Leave overnight at −20 °C.
9. Thaw the sample on ice, centrifuge for 2 min at a maximum speed. Transfer the supernatant to a new 2 mL tube and add 0.7 volume of 100% ethanol. Gently pipette.
10. Transfer the lysate to the RTA spin filter (maximum 800 μL).
11. Centrifuge for 2 min at 11,000 × *g*.
12. Discard the flow-through and replace the receiver tube.
13. Transfer the last part of the lysate to a spin filter and centrifuge as described above.
14. Wash the spin filter with 600 μL of Wash Buffer R1, centrifuge at 11,000 × *g* for 1 min, discard the flow-through.
15. Wash the spin filter with 700 μL of Wash Buffer R2, centrifuge at 11,000 × *g* for 1 min, discard the flow-through.
16. Repeat the washing step from **step 15**.
17. Centrifuge for 4 min at maximum speed to eliminate any residual ethanol.

16. Discard the RTA Receiver tube. Transfer the spin filter set to a new 1.5 mL tube and add 40 μL of nuclease-free water.

18. Wait for 2 min and centrifuge at 11,000 × *g* for 1 min.
19. Keep RNA on ice or store at −80 °C.

3.4 RNA Extraction from Plasma or Serum Using the mirVana™ miRNA Isolation Kit

1. Thaw the serum/plasma on ice.
2. Transfer 500 μL of serum/plasma to a new tube and add 600 μL of Lysis/Binding Solution and 60 μL of miRNA Homogenate Additive.
3. Vortex and keep on ice for 10 min.
4. Add 500 μl of Acid-Phenol:Chloroform and vortex for 60 s.

5. Centrifuge at room temperature for 15 min at a maximum speed.
6. Transfer the aqueous phase to a new tube and add 100% ethanol (1.25× volume of the aqueous phase). Vortex and pipette onto a filter cartridge, centrifuge for 15 s at 10,000 × *g*.
7. Wash the filter with 500 μL of Wash Solution 1.
8. Centrifuge for 10 s at 10,000 × *g*.
9. Wash the filter with 500 μL of Wash Solution 2/3.
10. Centrifuge for 10 s at 10,000 × *g*.
11. Repeat wash from **step 9**.
12. Preheat 100 μL of nuclease-free water up to 95 °C and pipette on the filter cartridge, centrifuge at a maximum speed for 30 s.
13. Keep RNA on ice or store at −80 °C

3.5 Quantity and Purity Control of the Obtained RNA

The purity of RNA is assessed using a spectrophotometer, e.g., NanoDrop, which analyzes the presence of RNA contamination with organic solutions and proteins. Keep RNA samples on ice.

1. Open the sampling arm and pipette 1 μL of nuclease-free water onto a lower measurement pedestal.
2. Perform the "Blank" measurement.
3. Measure the concentration of nuclease-free water, it should be on background level.
4. Open the sampling arm and pipette 1 μL of RNA onto lower measurement pedestal.
5. Close the sampling arm.
6. Measure RNA concentration at 260 nm.
7. Wipe the measurement pedestal once the sample measurement is completed.
8. Pure RNA shall have the 260/230 and 260/280 ratios close to 2.0 (*see* **Notes 1** and **2**).

3.6 Preparation of the Standard Curve (Serial Dilution of RNA)

The standard curve is prepared in order to accurately define the microRNA expression levels and estimate the potential differences between the analyzed samples, free from the bias resulting from differences in RNA concentration. The volume of each sample depends on the number of microRNAs that will be further analyzed in a qPCR reaction.

1. Keep RNA samples on ice (*see* **Note 3**).
2. Start with a high-purity RNA stock of a known concentration (e.g., 200 ng/μL).

3. Add 10 μL of stock RNA and 10 μL nuclease-free water to new tube (100 ng/μL).
4. Pipette gently.
5. Add 10 μL of the prepared RNA dilution and 10 μL nuclease-free water to a new tube (50 ng/μL).
6. Pipette gently.
7. Repeat **steps 5** and **6** twice, preparing the subsequent dilutions (25 ng/μL, 12.5 ng/μL).

3.7 Reverse Transcription (RT) Reaction of Known microRNAs Using the Commercially Available microRNA Analysis Assay

The assays for microRNA analysis are provided by numerous companies. If an assay amplifying the microRNA of choice is not available, *see* Subheading 3, **item 9**.

1. Allow all the components to thaw on ice.
2. Prepare the RT master mix to reverse-transcribe all the analyzed RNA samples together with the prepared standard curve (*see* **Notes 4** and **5**).
3. Prepare the RT master mix following the formula below (*see* **Notes 6–8**): The volume of each reagent shall be adjusted depending on the number of samples used for RT reaction. Add 10% of the volume to cover the potential pipetting losses

(a) 100 mM dNTPs	0.15 μL
(b) MultiScribe Reverse Transcriptase	1.00 μL
(c) 10× Reverse Transcription Buffer	1.50 μL
(d) RNase Inhibitor 20 U/μL	0.19 μL
(e) RT primer	3.00 μL
(f) Nuclease-free water	4.16 μL
Total	10.00 μL

4. Mix gently, and place the RT master mix on ice.
5. Dispense 10 μL of RT master mix into 0.2 mL reaction tubes and add 5 μL of RNA.
6. Each 15-μL RT reaction consists of 10 μL of the master mix and 5 μL of the RNA sample.
7. Use the following parameters to program the thermal cycler:

(a) 30 min	16 °C.
(b) 30 min	40 °C.
(c) 5 min	85 °C.
(d) 4 °C	∞.

8. Load the reaction tube into thermal cycler and start the reverse transcription reaction.

3.8 PCR Amplification of Known microRNAs Using the Commercially Available microRNA Analysis Assay

1. Keep the TaqMan probe protected from light (*see* **Notes 9** and **10**).
2. Allow the probe to thaw on ice and mix before using.
3. Prepare RT master mix following the example below (*see* **Notes 6** and **7**):

(a) TaqMan MicroRNA Assay (5×)	2.00 μL.
(b) The RT reaction product	1.00 μL.
(c) TaqMan 2× Universal PCR Master Mix	5.00 μL.
(d) Nuclease-free water	2.00 μL.
Total	10.00 μL.

4. It is recommended to perform at least three replicates of each reaction (*see* **Note 11**).
5. Dispense 10 μL of the PCR master mix to the reaction plate wells (*see* **Note 12**).
6. The PCR reaction plate should be sealed with an optical adhesive cover foil.
7. Use the following thermal cycling parameters:

(a) 10 min	95 °C.
(b) 15 s	95 °C.
(c) 60 s	60 °C (40 cycles).
(d) ∞	40 °C.

8. Load the reaction plate into the thermal cycler and start the PCR amplification (*see* **Note 13**).

3.9 Analysis of Novel microRNAs Using Custom-Designed Probes

MicroRNA quantification can be also performed using custom-designed primers and probes (Figs. 1 and 2) that can be synthesized in any biotechnological company. The method is based on [2] and facilitates the analysis of novel microRNAs, for which the commercial assays are not yet available.

3.9.1 Designing the Reverse Transcription Primers and Probes for qPCR

1. Verify the sequence of the microRNA of interest (*see* Figs. 1 and 2).
2. Identify the last six to eight nucleotides on the 3′ end of the mature microRNA.
3. Create a sequence which is a reverse-complement of the sequence identified in point 2.
4. Add the created sequence to the 3′ end of a universal sequence of the RT hairpin primer ("GTCGTATCCAGTGCAGG GTCCGAGGTATTCGCACTGGATACGAC"), creating the reverse transcription primer.
5. Check the sequence of the 5′ end of the mature microRNA.

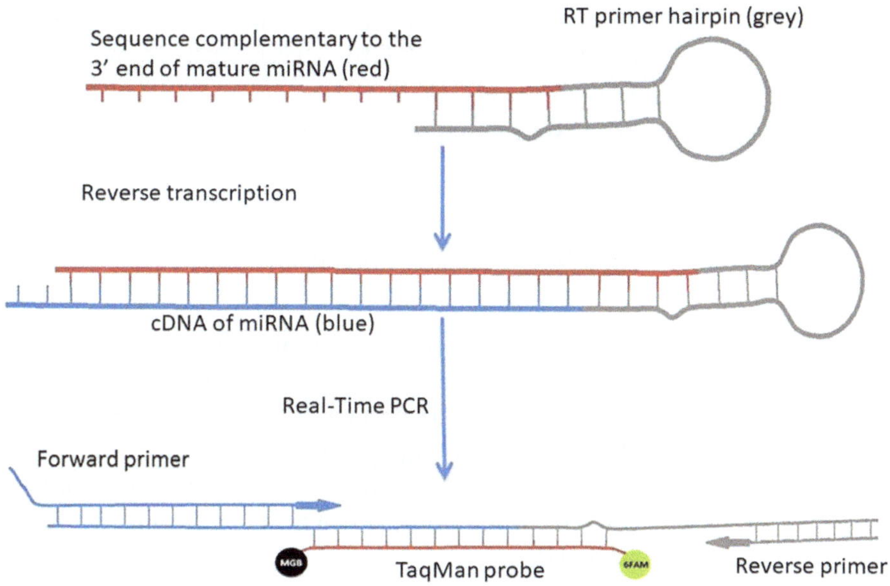

Fig. 1 Schematic presentation of the stem-loop primer design

UGAGAACUGAAUUCCAUGGGUU (5'end, 3' end)	**Sequence of the hsa-miR-146a-5p**
AACCCA	Reverse complement of the 3'end of miRNA
GTCGTATCCAGTGCAGGGTCCGAGGTATTCGCACTGGATACGAC	Generic RT hairpin primer
GTCGTATCCAGTGCAGGGTCCGAGGTATTCGCACTGGATACGAC+AACCCA	RT primer
TGAGAACTGAATTCCA	Sequence of the 5'end of miRNA
GCCGC	Generic forward qPCR primer
GCCGC+**TGAGAACTGAATTCCA**	Forward qPCR primer
GTGCAGGGTCCGAGGT	Middle part of the RT primer – reverse qPCR primer
TGGATACGAC+AACCCA	3' end of RT primer
6-FAM+TGGATACGAC+AACCCA+MGB	qPCR probe

Fig. 2 The stem-loop primer design on the example of hsa-miR-146a-5p (based on [8])

6. Create a sequence which is complement to the sequence identified in point 5.
7. Add the created sequence to the 3′ end of a universal sequence of the forward qPCR primer ("GCCGC"), creating the forward primer.
8. Use the middle part of the generic RT primer sequence ("GTGCAGGGTCCGAGGT") to create a universal reverse qPCR primer.

9. Confirm the specificity of primers with Primer-Blast Tool available on the NCBI server (https://www.ncbi.nlm.nih.gov/tools/primer-blast/):
   (a) Paste the sequence of the primers.
   (b) Set the PCR product size from 15 to 200.
   (c) Set "Genomes for selected organisms" as a database.
   (d) Choose the right organism.
   (e) Analyze the primers.
10. While ordering the primers and probes, add the 6-FAM and MGB to the 3′ end of the RT primer (underlined and red color) 6-FAM+ TGGATACGACAACCCA+MGB.
11. While ordering the primers and probes, order a synthetic DNA mimic of the analyzed miRNA. This will be necessary for setting the optimal reaction conditions (*see* Subheading 3, **item 10**).

3.10 Determining the Optimal Concentration of the RT Primer

Determination of the optimal concentration of the primer is necessary for sensitive and reliable reverse transcription and quantification of the analyzed microRNA (*see* **Note 14**).

1. Use a synthetic oligoDNA mimic of the analyzed microRNA—an oligonucleotide matching the microRNA sequence (*see* Subheading 3, **item 9**)
2. Dilute the oligoDNA according to the manufacturer's instructions.
3. Measure the concentration of thc oligoDNA solution.
4. Calculate the oligoDNA copy number in 1 μL of solution:
   (a) Use the ENDNEMO DNA/RNA Copy Number Calculator (http://www.endmemo.com/bio/dnacopynum.php).
   (b) Paste the sequence of oligoDNA.
   (c) Choose "ssDNA" from the list below the window.
   (d) Paste the oligoDNA concentration (measured in point 3) and click "Calculate."
5. Based on the obtained concentration, dilute the solution to receive a final concentration of 10^6 [copy number/μL]:
   (a) The concentration of oligoDNA is equal to the concentration received in point 4.
   [conc 1] = $a \times 10^b$.
   (b) Dilute the oligoDNA to receive the concentration [conc 2] = 1×10^c (the dilution shall result in dilution of the "*a*" number from pt. 4.1 to 1).

(c) Dilute the oligoDNA 10 times to receive the concentration [conc 3] = $1 \times 10^{c-1}$.

(d) Repeat **step 5c** to achieve a final concentration [conc] = 1×10^{6}.

6. Prepare the RT reaction mix following the formula below (*see* **Notes 6–8** and **15**):

(a)	RT buffer [5×]	2.00 μL
(b)	dNTP [100 mM]	0.12 μL
(c)	Reverse transcriptase [200 U/μL]	0.10 μL
(d)	RNase inhibitor [20 U/μL]	0.08 μL
(e)	RT primer of miRNA [tested concentration]	1.00 μL
(f)	Water	5.70 μL
(g)	oligoDNA of miRNA [10^6 copy number/μL]	1.00 μL
	Total	10.00 μL

7. The RT reaction should be performed under conditions determined by the assay manufacturer.

3.11 Validation of the qPCR Reaction Efficiency

Quantitative Real-Time PCR is performed using specific primers and probes designed based on the protocol presented above. The exact concentration of primers and probes has to be determined on the basis of the reaction efficiency.

1. The total volume of the reaction mix is 10 μL.
2. Each reaction should be performed in triplicates.
3. Prepare the following reaction mix (*see* **Notes 16** and **17**):

(a)	qPCR MasterMix 2×	5.00 μL.
(b)	Probe [tested concentration *XP*]	1.00 μL.
(c)	Forward primer [tested concentration *XF*]	1.00 μL.
(d)	Reverse primer [tested concentration *XR*]	1.00 μL.
(e)	Water	1.00 μL.
(f)	cDNA [from oligoDNA]	1.00 μL.

4. Set the qPCR reaction conditions as follows:

(a)	Preincubation	95 °C	10 min.
(b)	Amplification	50 cycles (95 °C—10s, 60 °C—30s, 72 °C—1s).	
(c)	Fluorescence measurement	72 °C.	
(d)	Cooling and incubation	40 °C.	

5. Run the reaction on a real-time PCR instrument. Our laboratory uses the LightCycler 480 II (Roche Diagnostics).

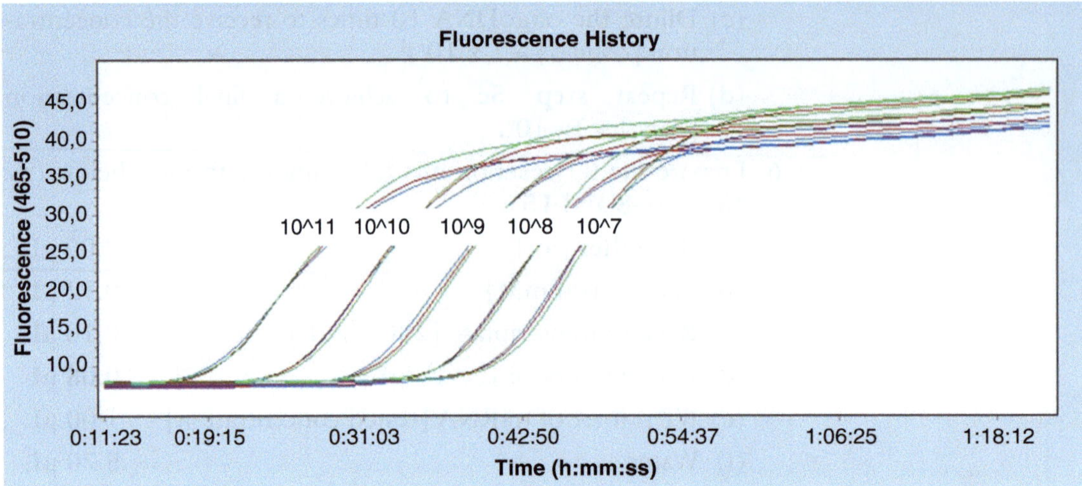

Fig. 3 qPCR amplification curves for different concentration of the synthetic microRNA mimic template

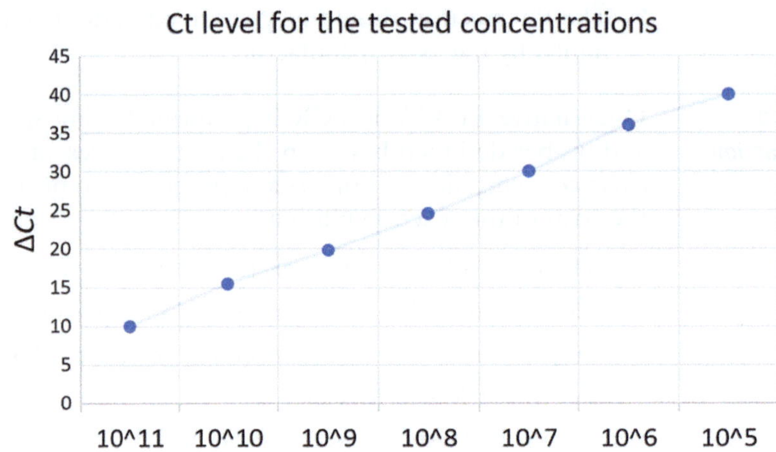

Fig. 4 Spot graph of the dilution curve, x = concentration of cDNA, y = Ct

6. The specificity of reaction is checked based on the absence of the amplification product in the negative control (non-template) reaction mix.
7. To determine the correct concentration of primers and probes, compare the reaction efficiency for novel miRNA to a known control miRNA. The optimal reaction mix is obtained when fluorescence curves for novel miRNA are similar to fluorescence curves for control miRNA.

3.12 Determine the Optimum RNA Concentration

1. Prepare the RT reaction mix as described in point 3.10–6 with different concentration of template (from 10^{11} to 10^{5} copy number of molecules), prepared as in point 3.10–5 (*see* **Note 18**).

2. Each Q-PCR reaction should be prepared in triplicate.
3. The optimal amount of RNA template in RT-PCR is determined on the basis of dilution curves (e.g., Fig. 3) and can be determined from the spot graph (Fig. 4) or calculated using the formula from the amplification curve (Fig. 5):

   $y = a^*\ln(x) + b$

   $x = e^{(\mathrm{Ct}-b)/a}$

   x—Copy number of molecules of RNA

   y—ΔCt

   a—Inclination of the trend line (speed of growth)

   b—Constant index (theoretical ΔCt value for $x = 0$) (*see* **Note 19**).
4. Calculate the obtained copy number to the concentration of RNA by the ENDNEMO DNA/RNA Copy Number Calculator (http://www.endmemo.com/bio/dnacopynum.php) as described in Subheading 3, **item 10** (*see* **Note 20**).

   When optimal condition of reaction is established, the analysis of expression of novel miRNA in tissue or cell line can be performed as described in Subheadings 3, **items 6–8**.

3.13 Data Analysis

The available analysis methods include absolute quantification and relative quantification (*see* **Notes 21–23**).

3.13.1 Absolute Quantification

1. Using the real-time PCR instrument software, calculate the Ct values for each sample (calculate the mean for each triplicate).
2. Prepare a graph with points: *Y* axis (Ct) and *X* axis (concentration) for each sample from serial dilution.
3. Generate a linear regression line, which represents the standard curve.
4. Calculate the line of trend pattern ($y = Ax + B$).
5. Comparing the Ct values of the unknown samples to this standard curve allows for quantification of concentration.
6. Calculate the concentration of unknown samples ($\Rightarrow x = (y - B)/A$).

3.13.2 Relative Quantification Normalized to the Reference Gene (the ΔCt Method)

The ΔCt method relies on assumption that the change of the Ct value for the normalizer (reference gene) has the same effect on relative quantity as the change in Ct for the unknown sample.

Example

	Test	Normalizer
Target miR	Ct (target, test)	Ct (target, norm)
Reference	Ct (reference, test)	Ct (reference, norm)

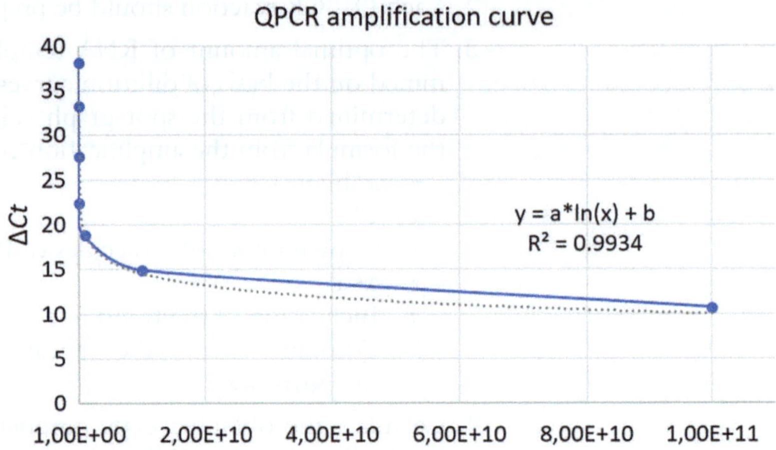

Fig. 5 Trend line of the dilution curve, x = concentration in copy number, y = Ct

This method uses the subtraction of reference and target Ct values for each sample.

1. Calculate the Ct values for each sample (a mean for a triplicate).
2. Calculate the ΔCt for the test sample and normalizer:
   ΔCt = Ct(reference)–Ct(target)
3. Calculate the relative expression for each sample:
   Sample expression = 2^ΔCt.
   Normalizer expression = 2^ΔCt.
4. Calculate the fold change expression:

   Fold change = sample expression/normalizer expression.

4 Notes

1. RNA purity is an important factor influencing the efficiency of the reverse transcription reaction. RNA purity can be assessed spectrophotometrically by determination of the absorbance at 260 and 280 nm (A260/A280). Pure RNA has the A260/A280 ratio of 1.9–2.1 in 10 mM Tris at pH 7.5.
2. Total RNA includes the full complement of RNAs: mRNA, miRNA, tRNA, and rRNA.
3. To prepare the standard curve, a stock sample with a known concentration is required. Examples of serial dilution:
   (a) 1:2 (400 ng, 200 ng, 100 ng, 50 ng, 25 ng, 12.5 ng; xμL RNA + xμL nuclease-free water);
   (b) 1:3 (200 ng, 67 ng, 22.2 ng, 7.4 ng, 2.5 ng; xμL RNA +2xμL nuclease-free water).

The sample volume depends on the number of microRNAs that will be analyzed in a qPCR reaction, including the reference gene.

4. It's recommended to use nondenatured RNA, free of RNase and free of substances that potentially inhibit the reverse transcription and PCR reaction. Denaturation may reduce the yield of cDNA synthesis. Working in an RNase-free environment is critical for performing successful cDNA synthesis reactions. Always wear gloves when handling the kit components and samples.
5. The main difficulty in quantification of miRNA results from the short length of the mature molecules (~22 nt). Currently the TaqMan Assay and quantitative real-time PCR are the methods, which assure unique sensitivity and specificity of miRNA detection and quantification. The TaqMan MicroRNA Assays include a target-specific stem-loop reverse transcription primer. The stem-loop structure included in the tail of the primer constitutes a key advantage of these assays and provides specific detection of the mature miRNA. The specific reverse transcription primer extends the 3′ end of the target to produce a template that can be used in real-time PCR.
6. When making a master mix for reactions (reverse transcriptase and PCR), increase the total volume by 10% to account for pipetting losses.
7. The given amounts of reagents are sufficient for a single analysis of 1 microRNA. The amount of the master mix should be adjusted based on the number of planned reactions (including the tested microRNA and normalizer).
8. It's possible to combine more than one RT primer in a single RT reaction, i.e., to multiplex the reverse transcription of more than 1 microRNA. To do this, adjust the water volume to the total reaction volume, e.g., for two RT primers use only 1.16 μL of the nuclease-free water. It is important to make sure that certain primer combinations do not interfere with each other and do not affect the PCR reaction. Thus, the expression levels of miRNAs amplified in a multiplex RT reaction should be initially compared to the levels produced when each miRNA is reverse transcribed alone.
9. It is recommended to use a separate laboratory zone for handling the qPCR reagents to minimalize the risk of contamination.
10. Taqman hydrolysis probes are considered the highest accuracy reagents for assessing expression of genes. These are dual labeled hydrolysis probes, which utilize the 5′ exonuclease activity of the Taq Polymerase for measuring the amount of target sequences in the samples. The 18–22 bp oligonucleotide long probe is labeled with a reporter fluorophore at the 5′ end and a quencher fluorophore at the 3′ end.

11. A negative control reaction should also be performed. Negative control reactions include a pure water sample and a sample that contains the reverse transcription mix without the template RNA. The absence of fluorescence in control samples shows the absence of DNA contamination and ensures high fidelity of the results.
12. Based on the manufacturer's protocol, the reaction shall be performed in 20 μL of the total volume, but the same results are obtained when a final volume of 10 μL is used.
13. PCR amplification must be performed using a Real-Time PCR system allowing for the detection and registration of the fluorescence signals. In this step, the polymerase amplifies target cDNA using sequence-specific primers and the amount of template miRNA is measured.
14. It is highly recommended to run the reaction for control miRNA using primers and probes also prepared according to the described method [2].
15. The recommended RT primer concentrations for the test include 1 μM, 2 μM, and 4 μM.
16. The recommended probe concentrations for the test include 1 μM, 2 μM, and 4 μM.
17. The recommended primer concentrations for the test include:

XF: 1 μM	1 μM	3 μM	5 μM	5 μM	7 μM	7 μM
XR: 1 μM	3 μM	1 μM	5 μM	7 μM	5 μM	7 μM

18. Use the concentration of primers and probes determined in point 2.10 and 2.11.
19. Trend line determinant (R^2) should be close to 1.
20. This time paste the copy number instead of concentration and click "Calculate."
21. Absolute quantification is obtained by comparing the Ct values of the tested sample to a standard curve. As a result, you obtain the quantity of nucleic acid (copy number, μg) per given amount of sample (per cell, per μg of total RNA). In a relative quantification, the obtained value represents the fold difference of the measured nucleic acid as compared to equivalent dilution of unknown and control sample. In both methods, the measurements obtained in a qPCR experiment must be normalized to a reference gene.
22. In absolute quantification, the results can be normalized to a known concentration (number of molecules) of a nucleic acid (synthetic oligoDNA) used as PCR template. In relative quantification, the results can be normalized to the expression levels of a reference gene. The approach is used to ensure that the analysis compares miRNA levels in equal amounts of sample material.

23. Selection of the proper endogenous control is the key to a proper data normalization. Correct reference gene is characterized by the stable expression level across all tested samples. Small non-coding RNAs (snoRNAs, snRNAs) are commonly used as reference genes in microRNA quantification (e.g., *RNU6B*, *RNU66*, *RNU43*). They show moderate but constant expression across a large number of tissues, so they are suitable candidates for an endogenous control. The analysis of miRNA expression in plasma is difficult due to lack of commercially available reference genes. Based on our experience, we suggest using miR-93-5p and miR-101-3p, which have been shown good and stable reference genes in plasma [9]. Also, a spike-in control of a known microRNA from a different species can be used. This can be obtained by adding known concentration of a synthetic microRNA to the plasma before microRNA extraction.

References

1. O'Connor L, Glynn B (2010) Recent advances in the development of nucleic acid diagnostics. Expert Rev Med Devices 7(4):529–539. https://doi.org/10.1586/erd.10.22
2. Chen C, Ridzon DA, Broomer AJ, Zhou Z, Lee DH, Nguyen JT et al (2005) Real-time quantification of microRNAs by stem-loop RT-PCR. Nucleic Acids Res 33(20):e179. https://doi.org/10.1093/nar/gni178
3. Kolanowska M, Wojcicka A, Kubiak A, Swierniak M, Kotlarek M, Maciag M et al (2017) Functional analysis of a novel, thyroglobulin-embedded microRNA gene deregulated in papillary thyroid carcinoma. Sci Rep 7(1):9942. https://doi.org/10.1038/s41598-017-10318-w
4. Lee LW, Zhang S, Etheridge A, Ma L, Martin D, Galas D et al (2010) Complexity of the microRNA repertoire revealed by next-generation sequencing. RNA 16(11):2170–2180. https://doi.org/10.1261/rna.2225110
5. Cloonan N, Wani S, Xu Q, Gu J, Lea K, Heater S et al (2011) MicroRNAs and their isomiRs function cooperatively to target common biological pathways. Genome Biol 12(12):R126. https://doi.org/10.1186/gb-2011-12-12-r126
6. Chugh P, Dittmer DP (2012) Potential pitfalls in microRNA profiling. Wiley Interdiscip Rev RNA 3(5):601–616. https://doi.org/10.1002/wrna.1120
7. Wojcicka A, Swierniak M, Kornasiewicz O, Gierlikowski W, Maciag M, Kolanowska M et al (2014) Next generation sequencing reveals microRNA isoforms in liver cirrhosis and hepatocellular carcinoma. Int J Biochem Cell Biol 53:208–217. https://doi.org/10.1016/j.biocel.2014.05.020
8. Jazdzewski K, Liyanarachchi S, Swierniak M, Pachucki J, Ringel MD, Jarzab B et al (2009) Polymorphic mature microRNAs from passenger strand of pre-miR-146a contribute to thyroid cancer. Proc Natl Acad Sci U S A 106(5):1502–1505. https://doi.org/10.1073/pnas.0812591106
9. Liu X, Zhang L, Cheng K, Wang X, Ren G, Xie P (2014) Identification of suitable plasma-based reference genes for miRNAome analysis of major depressive disorder. J Affect Disord 163(Supplement C):133–139. https://doi.org/10.1016/j.jad.2013.12.035

Chapter 8

MicroRNA Analysis Using Next-Generation Sequencing

Monika Kolanowska, Anna Kubiak, Krystian Jażdżewski, and Anna Wójcicka

Abstract

Next-generation sequencing (NGS) enables the analysis of both microRNA expression and sequence, allowing for elucidation of a comprehensive landscape of miRNAs in a given tissue and sample type. NGS analysis requires high-quality RNA extraction and preparation of microRNA libraries. In this chapter, we describe the methods used for RNA extraction from tissue specimens, serum, cytological slides, and formalin-fixed paraffin-embedded samples. Although the described library preparation and sequencing approaches are based on Illumina NextSeq 500 sequencing technology, the presented principles shall be compatible with other commercially available sequencing platforms.

Key words Next-generation sequencing, NGS, microRNA identification, microRNA quantification, RNA extraction, FFPE samples, Serum samples

1 Introduction

Next-generation sequencing brought new insight into the biology of microRNAs, allowing for simultaneous analysis of their expression and sequence variations [1]. Unlike other available methods—microarrays or quantitative real-time PCR, which require the use of pre-designed probes—NGS analysis allows for identification of all microRNA molecules present within the analyzed sample, thus it is not limited by the rapid discovery rate of novel microRNAs and novel sequence isoforms [2]. The method can be used for microRNA profiling and identification in tissue specimens, serum, cytological slides, or formalin-fixed paraffin-embedded samples [3, 4]. MiRNA analysis in NGS involves four important steps—miRNA/RNA extraction, preparation of libraries, RNA sequencing, and RNA sequence analysis. Novel library preparation kits do not require separation of miRNAs from total RNAs, allowing for a comprehensive miRNA analysis in limited amounts of the analyzed samples. RNA extraction method depends strictly on the analyzed material type. In this chapter, we provide our modified protocols

Ulf Andersson Vang Ørom (ed.), *miRNA Biogenesis: Methods and Protocols*, Methods in Molecular Biology, vol. 1823, https://doi.org/10.1007/978-1-4939-8624-8_8, © Springer Science+Business Media, LLC, part of Springer Nature 2018

of RNA extraction using commercially available reagents and kits. Preparation of libraries involves such manipulation of RNAs that they are compatible with the used sequencing system. Since the reads generated in miRNA analysis are by default short, the read length proposed by the sequencing platform is not a limitation. Current development in laboratory instrumentation significantly facilitated preparation of libraries as well as quantification and quality assessment of the obtained products.

2 Materials

Prepare all solutions using ultrapure water (prepared by purifying deionized water) and analytical grade reagents. Prepare and store all reagents at room temperature, unless otherwise indicated in the text.

2.1 Laboratory Facilities and Equipment

1. If possible, arrange separate laboratory rooms or laboratory space for pre- and post-PCR steps of the protocol to avoid RNA contamination.
2. Set of pipettes with adjustable dispensing volume:
   (a) 0.1–2.5 μL.
   (b) 2–20 μL.
   (c) 20–200 μL.
   (d) 100–1000 μL.
3. Filter tips for each volume.
4. Thermal cycler in the pre-PCR room and post-PCR room.
5. Quantus or Qubit for fluorescence-based nucleic acid quantification.
6. Spectrophotometer.
7. Thermoblock/water bath with a temperature range up to 85 °C.
8. Bioanalyzer or TapeStation Instrument (Agilent).
9. Pippin instrument (Sage Science).
10. Next-Generation Sequencer (in our protocol—NextSeq500, Illumina).
11. Centrifuge up to 20,000 × *g*.
12. Tissue homogenizer.
13. Mini spin centrifuge.
14. High-quality 1.5 and 2 mL tubes and PCR tubes.
15. Adhesive sealing foil for thermal cycler.
16. Hood.
17. Laboratory ice machine.

2.2 General Reagents for Molecular Biology

1. Xylene.
2. 100–96% ethanol.
3. Glycol.
4. 10 M NaOH.
5. 200 mM Tris–HCl, pH 7.
6. 20 mg/mL Proteinase K.
7. 0.1 M Dithiothreitol (DTT).
8. 3 M Sodium acetate.

2.3 Kits and Reagents for RNA Extraction and Analysis

1. RNA extraction from fresh/frozen tissue samples and cell cultures: TRIzol™ Reagent.
2. RNA extraction from formalin-fixed paraffin-embedded tissues and cytological slides: InviTrap Spin Universal RNA Mini Kit.
3. RNA extraction from plasma: mirVana™ miRNA Isolation Kit, with phenol.
4. RNA quantity and quality analysis: Agilent RNA 6000 Nano Kit; RNA QuantiFluor System.
5. DNA quantity and quality analysis: Agilent DNA High Sensitivity; dsDNA QuantiFluor System.
6. Kits for preparation of miRNAs library (stored at −20 °C): Multiplex Small RNA Library Prep Set; TruSeq Small RNA Library Preparation Kit.
7. NextSeq500 sequencing reagents: Buffer cartridge (stored at room temperature); NextSeq 500/550 High-Output v2 Kit (75 cycles)—reagent cartridge (stored at −20 °C); Accessory box containing hybridization buffer (HT1) (stored at −20 °C); High Output Flow Cell (stored at +4 °C).

3 Methods

Perform all the procedures at room temperature, unless otherwise indicated in the text.

3.1 RNA Extraction from Fresh/Frozen Tissue or Cell Cultures Using the TRIzol Reagent

1. In a single Eppendorf tube, add 750 μL of Trizol to 50–100 mg of tissue or to 10^5–10^7 monolayer cells and homogenize.
2. Incubate for 5 min, then add 400 μL of chloroform.
3. Mix by inverting the tubes 2–3 times, incubate for 5 min.
4. Centrifuge for 15 min at 12,000 × g at 4 °C.
5. Transfer the upper aqueous phase containing RNA to a new tube and add 400 μL of chloroform.
6. Mix by inverting the tubes 2–3 times, incubate for 5 min

7. Centrifuge for 15 min at 12,000 × *g* at 4 °C.
8. Transfer the upper aqueous phase to a new tube and add 600 μL of 96–100% ethanol with 0.3 M sodium acetate. Optionally store at −20 °C overnight.
9. Centrifuge for 10 min at 12,000 × *g* at 4 °C.
10. Discard the supernatant.
11. Wash the pellet in 1 mL of 75% ethanol and centrifuge for 5 min at 7500 × *g* at 4 °C.
12. Discard the supernatant and repeat wash.
13. Discard the supernatant, leave RNA on ice until it is dry.
14. Suspend RNA in nuclease-free water (*see* **Note 1**).
15. Keep RNA on ice or store at −80 °C.

3.2 RNA Extraction from Formalin-Fixed Paraffin-Embedded Tissue Samples Using the InviTrap Spin Universal RNA Mini Kit

1. Preheat the thermoblock/water bath to 48 °C.
2. Cut the FFPE block into 6–10 sections up to 10 μm and transfer to a 2 mL tube.
3. Add 0.5 mL of xylene. Mix until paraffin is dissolved.
4. Centrifuge for 2 min at a maximum speed.
5. Discard the supernatant and wash the pellet in 1 mL of 96–100% ethanol.
6. Centrifuge shortly at a maximum speed, discard ethanol, and leave the sample at 48 °C to evaporate.
7. Add 180 μL of RNase-free water, 40 μL of proteinase K (20 mg/mL), and 2.2 μL of 0.1 M DTT, and homogenize.
8. Incubate for 10 min at 48 °C and then heat to 80 °C for 10 min.
9. Add 900 μL of Lysis Solution TR (containing 0.01 M DTT) and ten Zirconia Beads II, vortex for 5 min. Transfer the supernatant to a new 2 mL tube. Leave overnight at −20 °C.
10. Thaw the samples on ice, centrifuge for 2 min at maximum speed. Transfer the supernatant to a new 2 mL tube, then add 0.7× volume of 100% ethanol. Gently pipette.
11. Transfer the lysate to the RTA spin filter (maximum 800 μL).
12. Centrifuge for 2 min at 11,000 × *g*.
13. Discard the flow-through and replace with a new receiver tube.
14. Transfer the last part of the lysate onto the RTA spin filter and centrifuge as described above.
15. Wash the spin filter in 600 μL of Wash Buffer R1, centrifuge at 11,000 × *g* for 1 min, discard the flow-through.

16. Wash the spin filter in 700 μL of Wash Buffer R2, centrifuge at 11,000 × *g* for 1 min, discard the flow-through, repeat the washing step.
17. Centrifuge the spin filter for 4 min at a maximum speed to eliminate any residual ethanol.
16. Discard the RTA Receiver tube. Transfer Spin filter set to a new 1.5 mL tube and add 40 μL of nuclease-free water (*see* **Note 1**).
18. Wait for 2 min and centrifuge at 11,000 × *g* for 1 min.
19. Keep RNA on ice or store at −80 °C.

3.3 RNA Extraction from Cytological Slides Using the InviTrap Spin Universal RNA Mini Kit

1. Soak the slide in xylene and mix under the fume hood until the cover glass is removed (from 2 h to overnight).
2. Preheat the thermoblock/water bath to 52 °C.
3. Remove the slide from xylene and scrape the tissue material off with a scalpel to a 1.5 mL tube.
4. Wash the pellet with 500 μL of 96% ethanol.
5. Centrifuge shortly, discard ethanol and leave the sample at 52 °C to evaporate.
6. Add 180 μL of RNase-free water, 40 μL of proteinase K (20 mg/mL) and 2.2 μL of 0.1 M DTT.
7. Incubate for 10 min at 48 °C and then for 10 min at 80 °C.
8. Add 900 μL of Lysis Solution TR (containing 0.01 M DTT), vortex for 5 min. Transfer the supernatant to a new 2 mL tube. Leave overnight at −20 °C.
9. Thaw the sample on ice, centrifuge for 2 min at a maximum speed. Transfer the supernatant to a new 2 mL tube and add 0.7 volume of 100% ethanol. Gently pipette.
10. Transfer the lysate to the RTA spin filter (maximum 800 μL).
11. Centrifuge for 2 min at 11,000 × *g*.
12. Discard the flow-through and replace the receiver tube.
13. Transfer the last part of the lysate to a spin filter and centrifuge as described above.
14. Wash the spin filter with 600 μL of Wash Buffer R1, centrifuge at 11,000 × *g* for 1 min, discard the flow-through.
15. Wash the spin filter with 700 μL of Wash Buffer R2, centrifuge at 11,000 × *g* for 1 min, discard the flow-through.
16. Repeat the washing step from **step 15**.
17. Centrifuge for 4 min at a maximum speed to eliminate any residual ethanol.

17. Discard the RTA Receiver tube. Transfer the spin filter set to a new 1.5 mL tube and add 40 μL of nuclease-free water (*see* **Note 1**).
18. Wait for 2 min and centrifuge at 11,000 × *g* for 1 min.
19. Keep RNA on ice or store at −80 °C.

3.4 RNA Extraction from Plasma or Serum Using the mirVana™ miRNA Isolation Kit

1. Thaw the serum/plasma on ice.
2. Transfer 500 μL of serum/plasma to a new tube and add 600 μL of Lysis/Binding Solution and 60 μL of miRNA Homogenate Additive.
3. Vortex and keep on ice for 10 min.
4. Add 500 μL of Acid-Phenol:Chloroform and vortex for 60 s.
5. Centrifuge at room temperature for 15 min at a maximum speed.
6. Transfer the aqueous phase to a new tube and add 100% ethanol (1.25× volume of the aqueous phase). Vortex and pipette onto a filter cartridge, centrifuge for 15 s at 10,000 × *g*.
7. Wash the filter with 500 μL of Wash Solution 1.
8. Centrifuge for 10 s at 10,000 × *g*.
9. Wash the filter with 500 μL of Wash Solution 2/3.
10. Centrifuge for 10 s at 10,000 × *g*.
11. Repeat wash.
12. Preheat 100 μL of nuclease-free water (*see* **Note 1**) up to 95 °C and pipette on the filter cartridge, centrifuge at maximum speed for 30 s.
13. Keep RNA on ice or store at −80 °C.

3.5 Quantity and Quality Control of the Obtained RNA

3.5.1 Quantity Analysis Using the Quantus Fluorometer

1. Place the RNA QuantiFluor System kit reagents on the bench to equilibrate to room temperature.
2. Prepare 1×TE buffer by diluting 20× TE buffer in nuclease-free water.
3. Prepare the 1× QuantiFluor RNA Dye working solution by diluting the dye in 1× TE buffer. The dilution depends on the used dye system.
4. Prepare 0.5 mL tubes; add 98 μL of TE buffer and 100 μL of the RNA Dye working solution.
5. Calibrate the Quantus instrument by measuring 2 μL of the RNA standard (100 ng/μL).
6. Add 2 μL of RNA to the tube prepared in **step 3**.
7. Perform the measurement (*see* **Note 2**).

3.5.2 Quality Analysis Using the Agilent 2100 Bioanalyzer

1. Place the Agilent RNA 6000 Nano reagents on the bench to equilibrate to room temperature.
2. Denature 2 μL of RNA ladder (provided by the manufacturer) and RNA samples for 2 min at 70 °C.
3. Keep the tubes on ice.
4. Add 550 μL of Agilent RNA 6000 Nano gel matrix to a spin filter.
5. Centrifuge for 10 min at 1500 × *g*.
6. Transfer 65 μL of the filtered gel to a new tube.
7. Add 1 μL of vortexed RNA 6000 Nano dye to 65 μL of the filtered gel.
8. Centrifuge for 10 min at 13,000 × *g*, RT.
9. Place the chip on the chip priming station and add 9 μL of the gel-dye mix to the well marked with a "G" symbol.
10. Close the chip priming station.
11. Press the plunger of the syringe down, wait for 30 s and then release the plunger.
12. After 5 s, slowly pull back the plunger to the 1 mL position.
13. Open the chip priming station and add 9 μL of the gel-dye mix to each of the wells marked.
14. Add 5 μL of the RNA 6000 Nano marker (green) into each of the 12 wells and to the well marked with the ladder symbol.
15. Pipette 1 μL of the RNA ladder to the well marked with the ladder symbol.
16. Pipette 1 μL of RNA to the sample wells.
17. Vortex the chip for 1 min and run on 2100 Bioanalyzer.
18. Verify RNA quality based on the obtained results (Fig. 1).

3.5.3 Purity Analysis Using the NanoDrop

The purity of RNA is assessed by a spectrophotometer, e.g., NanoDrop, which analyzes the presence of RNA contamination with organic solutions and proteins.

1. Keep RNA samples on ice.
2. Open the sampling arm and pipette 1 μL of nuclease-free water onto a lower measurement pedestal.
3. Perform the "Blank" measurement.
4. Measure the concentration of nuclease-free water, it should be on background level.
5. Open the sampling arm and pipette 1 μL of RNA onto lower measurement pedestal.
6. Close the sampling arm.
7. Measure RNA concentration at 260 nm.

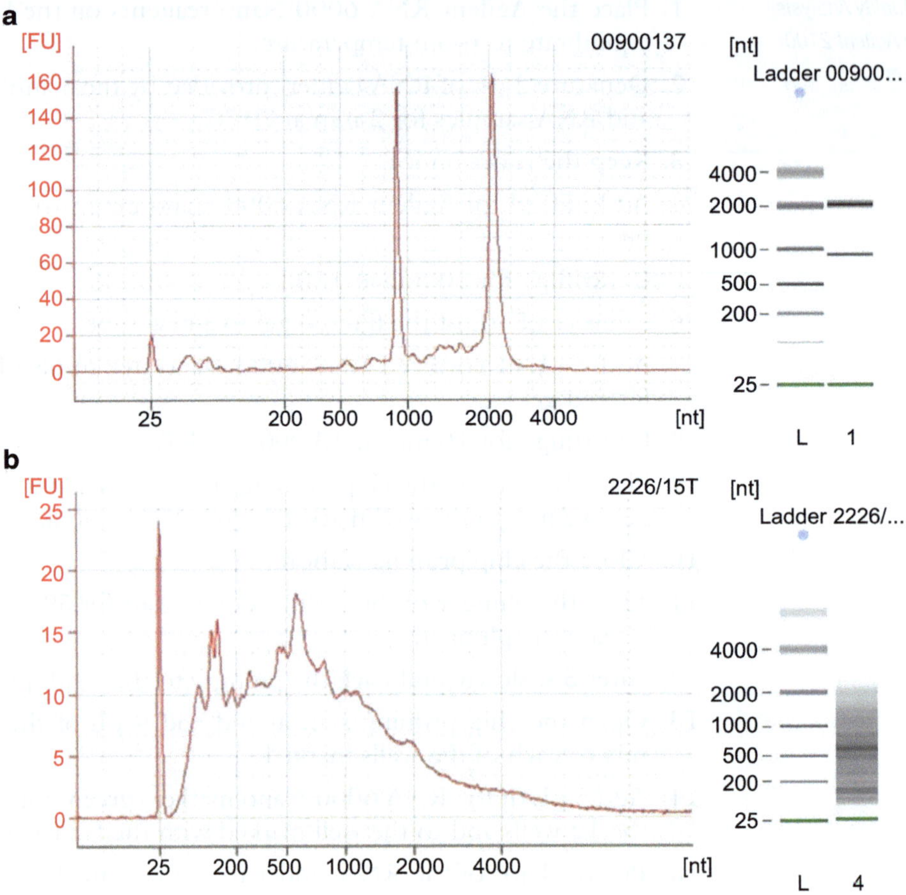

Fig. 1 RNA quality analysis on the 2100 Bioanalyzer. (**a**) High-quality RNA is characterized by two distinct peaks of 18 s and 28 s RNA (RIN = 10). (**b**) Degraded RNA is seen as a smear (RIN = 2.4)

8. Wipe the measurement pedestal once the measurement of sample is completed.
9. Pure RNA shall have the 260/230 and 260/280 ratios close to 2.0.

3.6 Ligation of Adapters

The miRNA sequencing assays provided by Illumina and New England Biolabs contain 48 single indexes, allowing for pooling and sequencing of up to 48 samples per one run.

3.6.1 3′ Adapter ligation

1. Use between 100 ng and 1 μg of input RNA in 5 μL of nuclease-free water (*see* **Note 3**).
2. Add 1 μL of the 3′ adapter to RNA and gently pipette (*see* **Note 4**).
3. Incubate the mix containing the 3′ adapter and RNA in 70 °C for 2 min and immediately put it on ice. This prevents from formation of secondary structures.

4. Preheat the thermal cycler to 28 °C.
5. Prepare the ligation mix on ice by adding:
   (a) Ligation Buffer 2 μL.
   (b) RNase Inhibitor 1 μL.
   (c) T4 RNA Ligase, Deletion Mutant 1 μL (*see* **Notes 5** and **6**).
6. Mix by pipetting.
7. Add the ligation mix (4 μL) to RNA and gently pipette.
8. Incubate at 28 °C for 1 h in a preheated thermal cycler.
9. Leave the tube in the thermal cycler and add 1 μL of the oligonucleotide blocking formation of adapter dimers (*see* **Note 7**).
10. Proceed with the 5′ adapter ligation as below.

3.6.2 5′ Adapter Ligation

1. Incubate the 5′ adapter in 70 °C for 2 min and put it immediately on ice. This prevents from formation of secondary structures.
2. Prepare the ligation mix on ice by adding:
   (a) 5′ adapter 1 μL.
   (b) T4 RNA Ligase 1 μL.
   (c) 10 mM ATP 1 μL.
3. Mix by pipetting.
4. Add the prepared ligation mix (3 μL) to RNA and gently pipette.
5. Incubate at 28 °C for 1 h in a preheated thermal cycler.
6. Place the mix on ice.

3.7 Reverse Transcription (RT)

1. Add 1 μL of RNA RT Primer to 6 μL of the mix obtained in Subheading 3.6.2.
2. Incubate the solution in 70 °C for 2 min.
3. Place the mix on ice.
4. Prepare the RT mix by adding:
   (a) 5× First Strand Buffer 2 μL.
   (b) 12.5 mM dNTP mix 0.5 μL.
   (c) 100 mM DTT 1 μL.
   (d) RNase Inhibitor 1 μL.
   (e) SuperScript II Reverse Transcriptase 1 μL.
5. Add 5.5 μL of RT mix to 7 μL of the RNA and RT primer mix.
6. Incubate in 50 °C for 1 h.
7. Place the mix on ice.

3.8 PCR Reaction

The library amplification allows for selective enrichment of cDNA fragments containing both 3′ and 5′adapters by using specific single barcoded primers complementary to the adapter sequences. Thus, only fragments containing the 3′ and 5′ adapters are amplified.

1. Prepare the PCR mix by adding:
   (a) Ultrapure Water 8.5 μL.
   (b) 2× PCR Master Mix 25 μL.
   (c) RNA PCR Primer 2 μL.
   (d) RNA PCR Primer Index 2 μL.
2. Combine the 37.5 μL of the PCR mix with 12.5 μL of cDNA.
3. Run the PCR reaction in the following conditions:
   (a) Set the lid temperature at 100 °C.
   (b) 98 °C for 30 s.
   (c) 11 cycles of:
       98 °C for 10 s.
       60 °C for 30 s.
       72 °C for 15 s.
   (d) 72 °C for 10 min.
   (e) +4 °C hold.

3.9 Library Quantification and Quality Check

The quality and quantity of the library must be carefully verified in automated electrophoresis, performed on Bioanalyzer, TapeStation (Agilent) or similar instrument. The peak corresponding to the microRNA-derived cDNA should range between 140 and 150 bp (Fig. 2).

3.9.1 Quantity Analysis Using the Quantus Fluorometer

1. Place the dsDNA QuantiFluor System kit reagents on the bench to equilibrate to room temperature.
2. Prepare 1×TE buffer by diluting 20× TE buffer in nuclease-free water.
3. Prepare the 1× QuantiFluor DNA Dye working solution by diluting the 200× concentrated dye in 1× TE buffer.
4. Prepare 0.5 mL tubes; add 98 μL of TE buffer and 100 μL of DNA Dye working solution.
5. Calibrate the Quantus by measuring 2 μL of DNA standard (100 ng/μL).
6. Add 2 μL of DNA to the tube prepared in **step 3** above.
7. Perform measurement.

3.9.2 Quality Analysis Using the Agilent 2100 Bioanalyzer

1. Pipette 9 μL of the gel-dye mix to the well marked with a "**G**" symbol. Press the plunger until it is held by the clip and wait for 1 min; then release the clip.
2. After 5 s slowly pull back the plunger to the 1 mL position.
3. Pipette 9 μL of the gel-dye mix to the well marked with a "**G**" symbol.
4. Pipette 5 μL of marker into each of the 11 wells and to the well marked with the ladder symbol.
5. Pipette 1 μL of DNA ladder to the well marked with the ladder symbol.
6. Pipette 1 μL of sample DNA to other wells.
7. Vortex the chip for 1 min.
8. Run the chip on the 2100 Bioanalyzer (Fig. 2).

3.10 Library Fractionation Using Pippin HT

The appropriate library fragments, containing microRNA-derived cDNA, can be separated through extraction from a 6% PolyAcrylamide gel. However, currently the most suitable option is the use of Pippin HT instrument (SageScience), which allows for remote selection of the desired library fragments.

1. Combine 20 μL of the cDNA library sample with 5 μL of the 30A solution (internal standard).
2. Vortex briefly.
3. On the PippinHT Software, choose the cassette: 3% Agarose, 90–250 bp.

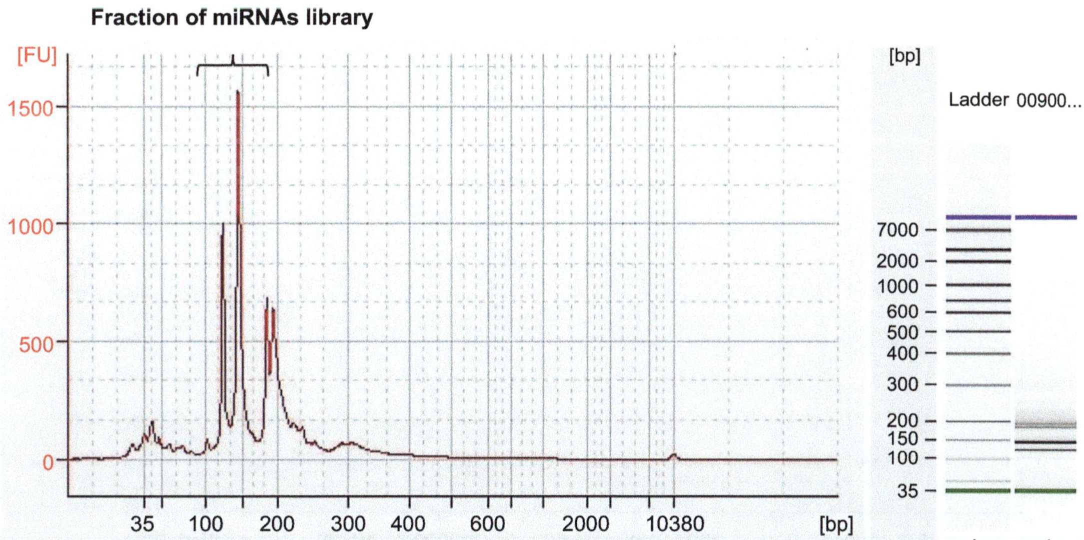

Fig. 2 Pre-separated library of small microRNAs has the length between 140 and 150 bp. Quality of DNA measured using the DNA High Sensitivity DNA Chip, 2100 Bioanalyzer. The amount and size of DNA is based on fluorescence unit [FU]

4. Enter the size selection parameters for each lane choosing 122–166 bp.
5. Inspect the gel columns on the chosen cassette. It is crucial to avoid any air bubbles in the cassette lanes.
6. Calibrate the optics of the device and prepare the cassette for loading, using calibration fixtures onto both optical nests.
7. Load samples to the cassette (25 μL of the combined cDNA library plus internal standard).
8. Start separation of cDNA.
9. Collect the size-separated libraries from elution wells (Fig. 3).

3.11 Pooling and Final Quality Check of the Obtained cDNA Fragments

Once the miRNA libraries have been obtained, they shall be concentrated before running the NGS sequencing. Pooling of the obtained libraries is possible thanks to the use of 48 different adapter sequences and allows for a significant reduction of the sequencing costs.

1. Add 2 μL of glycogen, 30 μL of 3 M NaOAc, and 975 μL of pre-chilled (−15° to −25 °C) 100% ethanol to 300 μL of extracted miRNA library.
2. Centrifuge at 20,000 × *g* for 20 min at 4 °C.
3. Remove the supernatant and wash the pellet with 500 μL of 70% ethanol (RT).
4. Centrifuge at 20,000 × *g* for 2 min (RT).

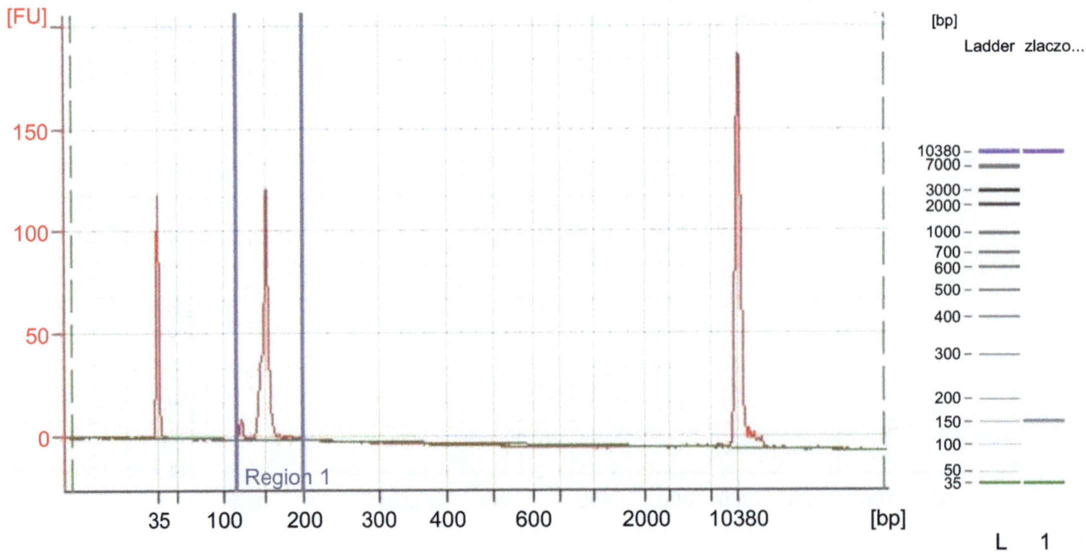

Fig. 3 The final miRNA library after separation on Pippin HT with one peak at 140–150 bp. See also lower marker (35 bp) and upper marker (10,380 bp)

5. Discard the supernatant and place the tube in a thermoblock set to 37 °C for 5–10 min or until the pellet is dry.
6. Resuspend the pellet in 10 μL of nuclease-free water.

Prior to sequencing, miRNA libraries should be pooled in equal DNA quantities. Verify DNA quality and quantity as described above (Fig. 3).

The final molar concentration of the library is calculated based on the concentration (in ng/μL) and average length of the library according to mathematical formula:

$$\text{Concentration in nM} = \frac{\text{Concentration}\left[\text{ng} / \mu L\right]}{\left(660\left[\text{g} / \text{mol}\right]\right) / \times \text{average library length}} \times 10^6$$

The library concentration used for sequencing depends on the sequencer, i.e., for NextSeq 500 and 550 it is recommended to use concentration between 1 and 4 nM, which is next diluted to 1.8 pM and prepared for sequencing. The sequencing of miRNA library should be performed as a single stranded nucleic acid analysis, with the shortest available length of read.

3.12 MicroRNA Sequencing on a NextSeq 500/550 Platform

1. Thaw all the necessary reagents on the bench at room temperature: NextSeq 500/550 High-Output v2 Kit (75 cycles)—reagent cartridge, accessory box containing hybridization buffer (HT1).
2. Remove the high output flow cell from +4 °C and allow it to equilibrate to room temperature.
3. Freshly prepare the dilution of 0.2 M NaOH from a 10 M NaOH stock for molecular biology.
4. Dilute the final library to 1 nM in nuclease-free water to obtain a final volume of 20 μL.
5. Add 20 μL of 0.2 M NaOH to denature the cDNA.
6. Mix gently by pipetting, incubate for 5 min at room temperature.
7. Add 20 μL of 200 mM Tris–HCl, pH 7.
8. Mix gently by pipetting.
9. Add 940 μL of a pre-chilled HT1 buffer to 60 μL of denatured cDNA mix to obtain 1 mL of total volume.
10. Keep the mix on ice.
11. In a new tube, add 117 μL of denatured library to 1183 μL of pre-chilled HT1 buffer.
12. Place the mix on ice.
13. Load the total volume of the mix (1300 μL) into the reagent cartridge.

14. Load the buffer cartridge (stored at room temperature), reagent cartridge, and high output flow cell into NextSeq.
15. Set the run parameters:
    (a) High output.
    (b) 75 cycles.
    (c) Single stranded.
    (d) Read 1—76 cycles.
    (e) Read 2—Index 1—7 cycles.
    (f) Read 3—Index 2—0 cycles.

4 Notes

1. RNA samples must be diluted in nuclease-free water, not DEPC-treated water.
2. The quality of RNA shall be verified on Bioanalyzer or TapeStation instrument and the purity shall be analyzed spectrophoretically. Currently the most accurate quantitation of nucleic acid concentration is achieved by the fluorescence binding dyes, while the spectrophotometer provides information related to the contamination of RNA with organic solutions and proteins, indicated by the 260/230 and 260/280 ratios below 1.8.
3. Poor quality of input RNA can cause low library yield.
4. If the amount of input RNA is close to 100 ng, the adapters should be diluted 1:2 and the number of PCR cycles should be increased.
5. Preparation of microRNA libraries for next-generation sequencing is performed with the use of commercially available kits provided by various biotechnological companies. The main attitude of novel approaches is that there is no need of separating small RNA from total RNA prior to library preparation. Usually the kits allow the use of between 100 ng and 1 μg of input RNA in 5 μL of nuclease-free water. Using total RNA as the input material allows to minimize the wastage during RNA fractioning. Prior to library preparation, the quality and quantity of input RNA should be verified on Bioanalyzer or TapeStation (Agilent Technologies) and the RNA Integrity Number (RIN) value should be ≥ 7.
6. The first step of library protocol is the ligation of a modified 3′ adapter to the 3′ hydroxyl group, specific for small RNAs due to Dicer cleavage activity. The ligation is catalyzed by the T4 RNA ligase deletion mutant T4Rnl2(1–249). The enzyme doesn't require ATP for ligation reaction; however, it needs a

pre-adenylated substrate. Therefore the first step of ligation, based on addition of the 3′ adapter, selects only small noncoding RNAs for further preparation of libraries.

7. While using different kits for preparation of miRNA libraries, remember about various methods of avoiding the adapter dimer formation. In the Illumina protocol, the blocking oligonucleotides are used. In the New England Biolabs protocol, reverse transcription primer hybridizes to ligated adapter, stopping formation of dimers.

References

1. Cloonan N, Wani S, Xu Q, Gu J, Lea K, Heater S et al (2011) MicroRNAs and their isomiRs function cooperatively to target common biological pathways. Genome Biol 12(12):R126. https://doi.org/10.1186/gb-2011-12-12-r126
2. Swierniak M, Wojcicka A, Czetwertynska M, Stachlewska E, Maciag M, Wiechno W et al (2013) In-depth characterization of the microRNA transcriptome in normal thyroid and papillary thyroid carcinoma. J Clin Endocrinol Metab 98(8):E1401–E1409. https://doi.org/10.1210/jc.2013-1214
3. Koperski L, Kotlarek M, Swierniak M, Kolanowska M, Kubiak A, Gornicka B et al (2017) Next-generation sequencing reveals microRNA markers of adrenocortical tumors malignancy. Oncotarget 8(30):49191–49200. https://doi.org/10.18632/oncotarget.16788
4. Tam S, de Borja R, Tsao MS, McPherson JD (2014) Robust global microRNA expression profiling using next-generation sequencing technologies. Lab Investig 94(3):350–358. https://doi.org/10.1038/labinvest.2013.157

Chapter 9

Identification of microRNA Precursor-Associated Proteins

Thomas Treiber, Nora Treiber, and Gunter Meister

Abstract

MicroRNA (miRNA) biogenesis is regulated intricately at multiple levels. In addition to transcriptional control of pri-miRNA loci, sequence as well as structural features of the pri-miRNA-stem loop determine its processing efficiency by the endonucleases Drosha and Dicer. On the one hand, general features are necessary to allow a hairpin to be recognized by the processing machinery; on the other hand, specific sequence motifs of individual miRNA precursors can be read by RNA binding proteins (RBPs) that regulate processing, leading to increased or decreased levels of functional miRNAs. In a pulldown experiment using the pri-miRNA hairpin as immobilized bait, cognate RBPs can be isolated and analyzed by immunoblotting or mass spectrometry, allowing for the discovery or analysis of protein regulators of miRNA biogenesis.

Key words RNA pulldown, Magnetic beads, microRNA precursor, RNA binding proteins

1 Introduction

Since their discovery more than two decades ago [1, 2], a tremendous amount of experimental studies have been performed and miRNAs are now recognized as fundamental gene regulators in almost all eukaryotes [3]. Hundreds of miRNAs are expressed in human cells and each cell type is characterized by a unique miRNA profile, helping to establish and maintain a distinct gene expression program by tuning target gene expression. Consequently, miRNA profiles and the corresponding gene expression programs are altered in diseases such as cancer [4]. Interfering with miRNA regulation is thus a promising and innovative direction for therapy [5].

The general principles of canonical miRNA biogenesis and function are reasonably well understood [6]. MiRNAs are organized as individual genes or gene clusters that are transcribed by RNA polymerase II. Alternatively, miRNAs can be located in introns and be transcribed together with the host gene or can be processed from other non-coding RNAs [3, 7–12]. In all cases, transcription generates a primary transcript, referred to as pri-miRNA, which

Ulf Andersson Vang Ørom (ed.), *miRNA Biogenesis: Methods and Protocols*, Methods in Molecular Biology, vol. 1823, https://doi.org/10.1007/978-1-4939-8624-8_9,

contains a stem-loop structure carrying 5′ and 3′ single-stranded flanking sequences. This structure is recognized by a complex composed of the RNase III Drosha and its co-factor DGCR8 (also known as the microprocessor) [13–16]. DGCR8 places Drosha on the pri-miRNA and Drosha cleaves at the basis of the stem, generating a miRNA precursor hairpin (pre-miRNA) [17]. This processing intermediate is transported to the cytoplasm using Exportin 5 as export receptor, where the RNase III Dicer cleaves the hairpin at the apical end of the stem and produces a short double-stranded RNA [18]. In subsequent steps, one strand is selected and incorporated into the RNA-induced silencing complex (RISC), where it associates with a member of the Argonaute (Ago) protein family [19–21]. The miRNA targets RISC to partially complementary sequences located in the 3′ untranslated region of target mRNAs. In subsequent steps, a member of the GW protein family is recruited (termed TNRC6A-C in mammals). These proteins are largely unstructured and potentially highly flexible proteins establishing protein–protein contacts to almost all factors acting downstream in gene silencing. For example, poly(A) binding proteins are contacted and deadenylation and decapping enzymes are recruited. The poly(A) tail is subsequently removed, translation inhibited, and the mRNA degraded [22].

As mentioned before, correct miRNA expression levels are critical for cell type-specific gene expression. Therefore, it is not astonishing that miRNA expression is tightly controlled. This is in part achieved by transcription factors such as p53 or myc [23–26]. However, miRNAs are also regulated post-transcriptionally on the pri- and pre-miRNA levels. For example, the RBP hnRNPA1 associates with pri-miR-18a and stimulates Drosha processing [27, 28]. Other examples are the RBPs TDP43 and YBX1, which can also affect miRNA expression on a post-transcriptional level [29, 30]. The most prominent example of cytoplasmic pre-miRNA regulation of the Dicer cleavage step is the stem cell-specific factor LIN28A. This RBP specifically binds to members of the let-7 miRNA family and recruits the TUTases TUT4/7, which add oligo(U) tails to the 3′ end of the pre-miRNA. Since Dicer recognizes the ends of the pre-miRNA, uridylation prevents Dicer binding and thus further processing. Instead, the 3′ to 5′ exoribonuclease DIS3L2 is recruited and the pre-miRNA is degraded [31–36].

In a recent large-scale proteomics study, we have identified a large number of RBPs that could affect miRNA biogenesis on a post-transcriptional level [37]. Interestingly, many of these factors are common RBPs involved in processes such as pre-mRNA splicing or coupling of transcription with downstream mRNA maturation steps. These observations suggest that the mRNA biogenesis machinery is tightly linked to pri-miRNA processing, which might be particularly important for the expression of intron-resident miRNA genes.

In the following chapter, we provide a detailed protocol for the identification of RBPs that specifically interact with pri- or pre-miRNAs. This experimental strategy can also be used for the detailed mapping and validation of RNA–protein interactions.

2 Materials

Special care should be taken when working with RNA. Although the short, structured RNAs used as baits in the pulldown reaction are not particularly sensitive to degradation by RNases, gloves should be worn during all handling procedures, filter tips are recommended for pipetting steps, and all buffers and solutions should be prepared using RNase-free water. Water of sufficient quality can be obtained by addition of 0.1% diethyl-pyrocarbonate (DEPC) followed by >2 h stirring and an autoclaving cycle to remove non-reacted DEPC.

2.1 RNA Bait Transcription and Purification

1. Template DNA.
2. 10× transcription buffer: 0.3 M Tris–HCl pH 8.0, 0.25 M $MgCl_2$, 0.1% (v/v) Triton X-100, 0.1 M DTT, 2 mM spermidine.
3. T7 RNA polymerase can be obtained from different manufacturers. We purify recombinant His-tagged T7 RNA polymerase from *E. Coli*. For commercial preparations, we estimate 1000 U = 1 μg.
4. NTP solution: 0.1 M ATP, CTP, GTP and UTP dissolved in DEPC-treated water. Adjust to pH 7 using 2 M Tris-base solution. Aliquot and store at −20 °C.
5. Thermostable inorganic pyrophosphatase.
6. Urea gel system.
7. RNA loading dye: formamide containing 0.01% (w/v) bromphenol blue.
8. TBE buffer: 89 mM Tris, 89 mM boric acid, 2 mM EDTA.
9. ALUGRAM Xtra SIL G/UV Thin layer chromatography (TLC) plates.
10. Ultra Low Range DNA Ladder.
11. Ethidium bromide staining solution: 1 μg/mL ethidium bromide in 1× TBE buffer.

2.2 Pulldown Experiment

1. Dynabeads Streptavidin M-270.
2. Adaptor oligonucleotide: 5′-AGGCUAGGUCUCCC-3′; all bases are 2′-*O*-methyl modified RNA. A biotin modification at the 3′-end is used to immobilize the oligonucleotide on the

streptavidin-coated Dynabeads. The adaptor oligo is dissolved in water at a concentration of 1 μg/μL and stored at −20 °C.

3. Pulldown buffer: 50 mM Tris–Cl pH 8.0, 150 mM NaCl, 5% (v/v) glycerol.
4. Stock solutions of 1 M dithiothreitol (DTT) and 0.1 M 4-(2-Aminoethyl)benzenesulfonyl fluoride hydrochloride (AEBSF).
5. Magnetic separation rack.
6. Branson Sonifier 450 Analog Ultrasonic Homogenizer.
7. Wash buffer I: 50 mM Tris–HCl pH 8.0, 300 mM NaCl, 5% (v/v) glycerol.
8. Wash buffer II: 50 mM Tris–HCl pH 8.0, 150 mM NaCl, 5% (v/v) glycerol, 0.1% (v/v) Triton X-100.
9. SDS sample buffer: 60 mM Tris–HCl pH 6.8, 50 mM DTT, 10% (v/v) glycerol, 2% (w/v) SDS, 0.02% (w/v) bromphenol blue.

3 Methods

To identify interactors of a bait RNA hairpin, in vitro transcribed RNA is coupled by a biotinylated adaptor to magnetic streptavidin beads. This bait is incubated with a cell lysate that has undergone a preclearing step. The beads are washed with three different buffers (Fig. 1). Bound proteins arc eluted with SDS loading buffer and analyzed by SDS-PAGE and western blot or mass spectrometry analysis. Examples of western blot analyses of pulldown experiments with either overexpressed or endogenous protein are shown in Fig. 2.

3.1 Template Design and Generation

1. As RNA bait for the pulldown experiment, the complete stem-loop structure contained in the pri-miRNA is chosen, as this is expected to contain most of the regulatory elements. Extensive single-stranded flanking sequences should be avoided as this leads to less stable baits and higher background binding (*see* **Note 1**).
2. The RNA bait is transcribed using T7 polymerase, which requires a double-stranded DNA template containing the promoter sequence 5′-TAATACGACTCACTATAGGG-3′. Transcription starts with the first G of the terminal triplet leading to an RNA starting with GGG. For immobilization of the RNA, a 5′-leader sequence is added to the hairpin of interest. Including the three guanines encoded by the T7 promoter its sequence is 5′-GGGAGACCTAGCCT-3′.

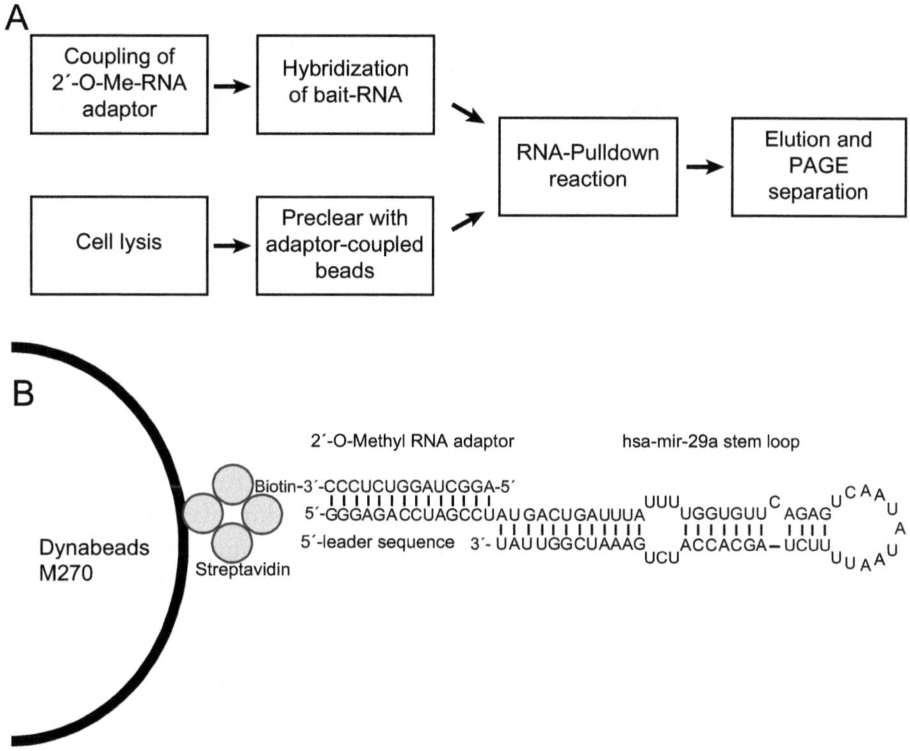

Fig. 1 Overview of the pulldown experiment and RNA-tethering assembly. (**a**) Flowchart of a miRNA-hairpin pulldown experiment. (**b**) Schematic representation of the immobilized pri-miRNA hairpin of human mir-29a. The different parts of the RNA assembly are labeled, 5′ and 3′ ends are indicated. Base pair interactions are shown as connecting lines

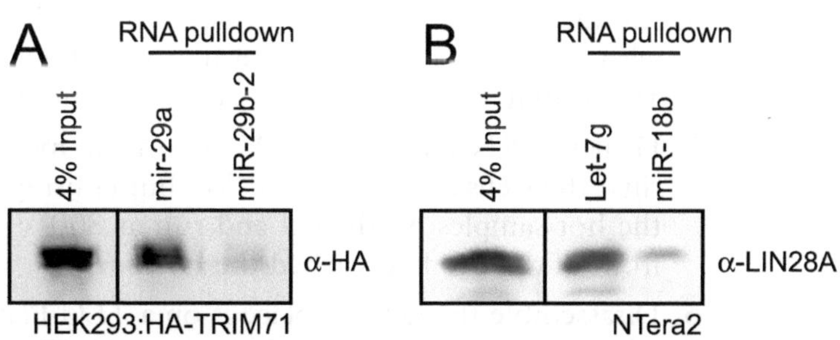

Fig. 2 Pulldown of overexpressed and endogenous RNA binding proteins. (**a**) RNA pulldown of TRIM71 protein overexpressed as HA-fusion protein in HEK293 cells. 4% of the input as well as the elution from pulldown experiments with hsa-miR-29a and hsa-miR-29b-2 were analyzed by SDS-PAGE followed by immunoblot analysis using an anti-HA antibody. (**b**) RNA pulldown with the hsa-let-7g and hsa-mir-18b hairpins from NTera2 cell lysate. 4% of the input and the pulldown elution fractions were analyzed by immunoblot against LIN28A

3. miRNA hairpins are usually encoded as undisrupted sequences in the genome and can therefore be amplified from genomic DNA by PCR. An extension containing the T7 promoter and the leader sequence can be added to the 5′-PCR primer (5′-TAATACGACTCACTATAGGGAGACCTAGCCT-3′). We recommend to clone the PCR product into a vector in order to verify the identity of the amplicon by sequencing before proceeding to the transcription step. Furthermore, the yield of the PCR reaction from a vector is generally higher than from genomic DNA (*see* **Note 2**).
4. As template for the transcription reaction, the miRNA-hairpin sequence with the 5′-extension containing the T7 promoter and RNA-leader sequences is amplified by PCR from the sequence-verified plasmid template and purified via a 2–3% (w/v) agarose gel (*see* **Note 3**). After extraction from the gel using a commercially available kit (we use NucleoSpin® Gel and PCR Clean-up, Macherey-Nagel), the template DNA is eluted in 30 μL of DEPC-treated water and the concentration is measured.

3.2 RNA Bait Transcription and Purification

1. 2 μg of DNA template are used in a 1 mL in vitro transcription reaction containing 100 μL of 10× transcription buffer, 50 μL NTP mix (0.1 M each), 50 μL T7 polymerase (2 mg/mL), and 2 μL thermostable inorganic pyrophosphatase. Mix gently and incubate at 37 °C for 3–16 h (*see* **Note 4**).
2. To stop the transcription, add 1 mL of RNA loading dye, mix and heat at 95 °C for 3–5 min (*see* **Note 5**).
3. Prepare a 10% urea PAGE gel with the dimensions 180 × 120 × 2 mm containing two wells 70 mm wide.
4. Mount the gel into a running chamber and fill the buffer tanks with 1× TBE buffer. Pre-warm the gel by running it for 15–30 min at 300 V before loading the samples.
5. Heat the RNA-samples to 95 °C; in the meantime, carefully rinse the wells of the gel with TBE buffer using a syringe. Load the hot samples on the gel and run at 300 V until the upper front of the dye has reached the bottom.
6. Disassemble the gel and place it on a TLC-Plate covered with a layer of saran wrap. Illumination of the TLC-plate with UV light (254 nm) results in a bright green fluorescence. Regions of the gel containing RNA are visible as dark shadows on the TLC-plate and are excised using a clean steel blade. If a lane shows more than one strong band, all bands are excised and processed separately.
7. Grind the gel slices into small pieces between two clean glass plates and collect the resulting material in a 15 mL reaction tube. Pipet DEPC-treated water into the tube until all gel

pieces are completely covered, then add another 1.5 mL and rotate the samples at 4 °C for at least 4 h.

8. Centrifuge the gel elution reactions for 10 min at 3000 × *g* and transfer 2 × 650 μL into two 1.5 mL reaction tubes. Add 50 μL 5 M NaCl and 600 μL 2-propanol to precipitate the RNA. Mix thoroughly and place at −20 °C for at least 1 h. Centrifuge for 20 min at 20,000 × *g* and 4 °C and discard the supernatant. Wash the pellet once with 1 mL ice-cold 80% (v/v) ethanol in DEPC-treated water and centrifuge again for 10 min at 20,000 × *g*.
9. Aspirate the supernatant with a 1000 μL pipet tip, let the residual supernatant settle for 2 min, then aspirate it with a 200 μL tip. Place the open reaction tube in a 55 °C heat block and dry the pellet for 5 min. Add 50 μL DEPC-treated water and shake the samples at 55 °C for 5 min. Pool the two elution reactions of the same bands and measure the concentration of the purified RNA.
10. Purity and integrity of an RNA bait are checked by analyzing 2 μg of RNA on a 10% urea PAGE gel. We use GeneRuler Ultra Low Range DNA Ladder supplemented with RNA loading dye as a marker to verify the size of the RNA (*see* **Note 6**). The gel is stained with 1 μg/mL ethidium bromide solution. If more than one band has been cut out of the preparative gel under Subheading 3.2, **step 6**, the right band can be identified from this gel (*see* **Note 7**).

3.3 Preparation of RNA-Coupled Beads

1. Thoroughly shake the Dynabeads suspension and transfer 100 μL per pulldown reaction into a 1.5 mL reaction tube (*see* **Note 8**). Collect the beads using the magnetic separation rack and aspirate the supernatant. Wash the beads twice with 1 mL pulldown buffer.
2. Resuspend the washed beads in an appropriate volume of pulldown buffer and add 4 μg of adaptor oligo per pulldown sample. Incubate for >1 h at 4 °C under constant agitation.
3. Collect the beads using the magnetic separation rack and aspirate the supernatant. Wash the beads twice with 1 mL pulldown buffer. Resuspend the beads in the original suspension volume (100 μL per pulldown sample).
4. For each sample, transfer 50 μL of the beads suspension into a new reaction tube. Add 10 μg of bait-RNA and adjust the volume to 200 μL with pulldown buffer. Incubate the coupling reaction >2 h at 4 °C under constant agitation. Store the rest of the beads at 4 °C (*see* **Note 9**).
5. Collect the beads as above and wash twice with pulldown buffer. Resuspend the RNA-coupled beads in 50 μL pulldown buffer per sample (*see* **Note 10**).

3.4 Extract Preparation and Pulldown Reaction

1. Harvest sufficient cells to obtain 10 mg of total protein for each binding reaction (*see* **Note 11**). Wash the cells once with PBS and resuspend the cell pellet in 1 mL of pulldown buffer supplemented with 1 mM DTT and 1 mM of the protease inhibitor AEBSF. Disrupt the cells with 20 pulses of sonication, cool on ice for 5 min, and apply another 20 pulses (*see* **Note 12**).
2. Centrifuge the lysate at 20,000 × *g* and 4 °C for 15 min to pellet insoluble material. Transfer the supernatant to a new tube.
3. Add 50 μL of the adaptor-coupled beads that were removed in Subheading 3.3, **step 4** to the cell lysate and mix carefully. Slowly rotate the samples at 4 °C for 2–6 h. After the incubation, centrifuge at 20,000 × *g*, 4 °C for 10 min to pellet the beads and any precipitated material (*see* **Note 13**).
4. Transfer the supernatant into a new tube and—if applicable—remove a sample as input for immunoblot analysis. Then add the RNA-coupled beads and rotate at least 2 h at 4 °C (*see* **Note 14**).
5. Collect the beads using the magnetic separation rack. If desired, remove a sample of the supernatant for immunoblot analysis. Aspirate the remaining supernatant and wash the beads once with wash buffer I, II and pulldown buffer, all supplemented with 1 mM DTT and 1 mM AEBSF, in this order. After the third wash, make sure to remove all supernatant and elute the bound proteins in 35 μL loading dye, e.g., NuPAGE LDS Sample Buffer with NuPAGE Sample Reducing Agent (Thermo Fisher Scientific) (*see* **Notes 15** and **16**). Heat the samples to 95 °C for 5 min, then pellet the beads by brief centrifugation. The samples are run on NuPAGE gels (4–12% Bis-Tris-Gel gradient gels, 1.0 mm) and further processed for mass spectrometry analysis.

4 Notes

1. MicroRNA hairpin sequences are annotated in the web resource miRBase (www.mirbase.org) under the subitem "stem-loop." The "get sequence" button leads to a linear sequence view. However, the choice of sequence termini is not very consistent (and might be worse for less common miRNAs or organisms) and should be checked carefully.
2. Many common cloning vectors contain a T7 promoter, which is also the annealing site for the T7 sequencing primer. If this is the case for your vector of choice, a different primer has to be used for sequencing and the vector cannot be used as a run-off transcription template.

3. Alternatively, a linearized plasmid can be used in a run-off transcription. If this is desired, the amount of template has to be increased according to the size of the plasmid in relation to the part that should be transcribed. Furthermore, the vector must not contain an additional T7 promotor (see also **Note 2**).
4. The optimal incubation time for a given transcription reaction can be optimized. Usually, the maximum yield is reached after 4 h. Incubation overnight can be very convenient and often results in high yields. Longer incubation times, however, increase the risk of RNA degradation, and it may be useful to add RNase inhibitor to overnight transcriptions.
5. The formamide in the RNA loading dye inhibits RNases possibly contained in the sample. At this stage, the RNAs are stable and can be stored for a longer period of time.
6. As the GeneRuler Ultra Low Range Ladder is a DNA marker, RNA bands run mostly a littler higher compared to the marker. Furthermore the gel should not get too hot, because otherwise the double-stranded DNA marker starts to denature, leading to bands that are more difficult to interpret.
7. Only highly pure and intact RNA of the right size should be used for the pulldown experiment. However, as ethidium bromide detection is very sensitive and the RNA is not stained proportional in higher amounts, the quality of the RNA preparations seems often worse than it is. We recommend to use RNAs for which the main band amounts to about 80–90% of the total stained RNA.
8. When working with overexpressed proteins, e.g., carrying tags, lower amounts of beads can be used to save costs. We apply 50 μL of beads for a pulldown reaction with Flag-HA-tagged proteins, using 20 μL for the preclear step and 30 μL for the pulldown reaction.
9. The incubation of the RNA coupling reaction can be extended overnight, which can be helpful especially with high sample numbers.
10. In order to achieve a high coupling density, the bait-RNA is used in excess and more than half of the RNA input is contained in the supernatant of the coupling reaction. If expensive synthetic RNA or material from low-yield transcription reactions is used, the unbound RNA can be precipitated and reused.
11. The amount of cells needed varies widely across different cell lines and may require a pilot experiment to estimate the amount of protein obtained per cell culture vessel. For this, i.e. two confluent plates of cells are harvested and lysed in 1 mL of lysis buffer as described above. After centrifugation, the protein content of the lysate is measured using a Bradford assay or similar method. For immunoblot-based readouts, the amount

of required sample is, of course, related to the strength of the detected input signal. In this case, the protein of interest should be clearly visible in 2–5% of the input material.

12. The sonication can be performed for each sample individually or with pooled samples of the same kind in a larger volume. The settings for the sonication depend on the volume and viscosity of the sample. We generally work with 50% duty cycle and use a microtip for volumes below 5 mL. Typically, the optimal power level for a 1 mL sample is between 2 and 3, but has to be adjusted such that the sample shows agitation and formation of small internal bubbles. Extensive foaming should be avoided. The cell lysis can be checked under a normal cell culture microscope. After sonication, no intact cells or nuclei should be visible. If necessary, an additional sonication cycle of 20 pulses can be applied. The sample should always be kept on ice during sonication.
13. This preclearing step reduces the background binding of proteins to the beads in the following pulldown experiment. It is highly recommended for experiments with mass spectrometry readout and greatly improves the results for some proteins in immunoblot. It may not be necessary for all proteins.
14. The pulldown reaction can be extended overnight, which is convenient in case of high sample numbers.
15. We use self-cast gels and the SDS sample buffer given in the materials section for western blot analysis. For mass spectrometry, commercially available precast gels are a better choice. These require a special loading dye that should be used instead of the SDS sample buffer in this case.
16. When setting up a new experiment, it is recommended to check the coupling efficiency and stability of the bait-RNA. For this, the pulldown experiment is performed in parallel with and without cell extract. After washing the beads, the elution is performed with 50% formamide containing 0.005% bromphenol blue. The eluate can be directly loaded onto a urea PAGE gel. 1 μg of the input bait-RNA and a size marker should also be loaded. After ethidium bromide staining of the gel, the coupling and integrity of the RNA can be evaluated. In case of problems with bait-RNA degradation, shorter incubation times and the addition of RNase inhibitors to the cell lysate are recommended.

Acknowledgements

Our research is supported by grants from the Deutsche Forschungsgemeinschaft (SFB 960, FOR2127, PP 1935, PP 1784) and the Bavarian Ministry for Education and Science (BioSysNet).

References

1. Wightman B, Ha I, Ruvkun G (1993) Posttranscriptional regulation of the heterochronic gene *lin-14* by *lin-4* mediates temporal pattern formation in *C. elegans*. Cell 75(5):855–862
2. Lee RC, Feinbaum RL, Ambros V (1993) The C. elegans heterochronic gene lin-4 encodes small RNAs with antisense complementarity to lin-14. Cell 75(5):843–854
3. Bartel DP (2009) MicroRNAs: target recognition and regulatory functions. Cell 136(2):215–233. S0092-8674(09)00008-7 [pii]. https://doi.org/10.1016/j.cell.2009.01.002
4. Garzon R, Calin GA, Croce CM (2009) MicroRNAs in cancer. Annu Rev Med 60:167–179. https://doi.org/10.1146/annurev.med.59.053006.104707
5. Ling H, Fabbri M, Calin GA (2013) MicroRNAs and other non-coding RNAs as targets for anticancer drug development. Nat Rev Drug Discov 12(11):847–865. https://doi.org/10.1038/nrd4140
6. Ha M, Kim VN (2014) Regulation of microRNA biogenesis. Nat Rev Mol Cell Biol 15(8):509–524. https://doi.org/10.1038/nrm3838
7. Ender C, Krek A, Friedlander MR, Beitzinger M, Weinmann L, Chen W, Pfeffer S, Rajewsky N, Meister G (2008) A human snoRNA with MicroRNA-like functions. Mol Cell 32(4):519–528
8. Scott MS, Ono M (2011) From snoRNA to miRNA: dual function regulatory non-coding RNAs. Biochimie 93:1987. S0300-9084(11)00183-0 [pii]. https://doi.org/10.1016/j.biochi.2011.05.026
9. Taft RJ, Glazov EA, Lassmann T, Hayashizaki Y, Carninci P, Mattick JS (2009) Small RNAs derived from snoRNAs. RNA 15(7):1233–1240. rna.1528909 [pii]. https://doi.org/10.1261/rna.1528909
10. Hasler D, Lehmann G, Murakawa Y, Klironomos F, Jakob L, Grasser FA, Rajewsky N, Landthaler M, Meister G (2016) The lupus autoantigen La prevents Mis-channeling of tRNA fragments into the human MicroRNA pathway. Mol Cell 63(1):110–124. https://doi.org/10.1016/j.molcel.2016.05.026
11. Okamura K, Hagen JW, Duan H, Tyler DM, Lai EC (2007) The Mirtron pathway generates microRNA-class regulatory RNAs in Drosophila. Cell 130(1):89–100
12. Berezikov E, Chung WJ, Willis J, Cuppen E, Lai EC (2007) Mammalian mirtron genes. Mol Cell 28(2):328–336
13. Landthaler M, Yalcin A, Tuschl T (2004) The human DiGeorge syndrome critical region gene 8 and its D. melanogaster homolog are required for miRNA biogenesis. Curr Biol 14(23):2162–2167
14. Han J, Lee Y, Yeom KH, Kim YK, Jin H, Kim VN (2004) The Drosha-DGCR8 complex in primary microRNA processing. Genes Dev 18(24):3016–3027
15. Denli AM, Tops BB, Plasterk RH, Ketting RF, Hannon GJ (2004) Processing of primary microRNAs by the microprocessor complex. Nature 432(7014):231–235
16. Gregory RI, Yan KP, Amuthan G, Chendrimada T, Doratotaj B, Cooch N, Shiekhattar R (2004) The microprocessor complex mediates the genesis of microRNAs. Nature 432(7014):235–240
17. Nguyen TA, Jo MH, Choi YG, Park J, Kwon SC, Hohng S, Kim VN, Woo JS (2015) Functional anatomy of the human microprocessor. Cell 161(6):1374–1387. https://doi.org/10.1016/j.cell.2015.05.010
18. Lund E, Guttinger S, Calado A, Dahlberg JE, Kutay U (2004) Nuclear export of microRNA precursors. Science 303(5654):95–98
19. Meister G (2013) Argonaute proteins: functional insights and emerging roles. Nat Rev Genet 14(7):447–459. nrg3462 [pii]. https://doi.org/10.1038/nrg3462
20. Ketting RF, Fischer SE, Bernstein E, Sijen T, Hannon GJ, Plasterk RH (2001) Dicer functions in RNA interference and in synthesis of small RNA involved in developmental timing in C. elegans. Genes Dev 15(20):2654–2659
21. Grishok A, Pasquinelli AE, Conte D, Li N, Parrish S, Ha I, Baillie DL, Fire A, Ruvkun G, Mello CC (2001) Genes and mechanisms related to RNA interference regulate expression of the small temporal RNAs that control *C. elegans* developmental timing. Cell 106:23–34
22. Jonas S, Izaurralde E (2015) Towards a molecular understanding of microRNA-mediated gene silencing. Nat Rev Genet 16(7):421–433. https://doi.org/10.1038/nrg3965
23. He L, He X, Lim LP, de Stanchina E, Xuan Z, Liang Y, Xue W, Zender L, Magnus J, Ridzon D, Jackson AL, Linsley PS, Chen C, Lowe SW, Cleary MA, Hannon GJ (2007) A microRNA component of the p53 tumour suppressor network. Nature 447(7148):1130–1134. nature05939 [pii]. https://doi.org/10.1038/nature05939
24. O'Donnell KA, Wentzel EA, Zeller KI, Dang CV, Mendell JT (2005) C-Myc-regulated

microRNAs modulate E2F1 expression. Nature 435(7043):839–843. https://doi.org/10.1038/nature03677

25. Tarasov V, Jung P, Verdoodt B, Lodygin D, Epanchintsev A, Menssen A, Meister G, Hermeking H (2007) Differential regulation of microRNAs by p53 revealed by massively parallel sequencing: miR-34a is a p53 target that induces apoptosis and G1-arrest. Cell Cycle 6(13):1586–1593
26. Raver-Shapira N, Marciano E, Meiri E, Spector Y, Rosenfeld N, Moskovits N, Bentwich Z, Oren M (2007) Transcriptional activation of miR-34a contributes to p53-mediated apoptosis. Mol Cell 26(5):731–743. S1097-2765(07)00318-8 [pii]. https://doi.org/10.1016/j.molcel.2007.05.017
27. Guil S, Caceres JF (2007) The multifunctional RNA-binding protein hnRNP A1 is required for processing of miR-18a. Nat Struct Mol Biol 14(7):591–596. nsmb1250 [pii]. https://doi.org/10.1038/nsmb1250
28. Michlewski G, Guil S, Caceres JF (2010) Stimulation of pri-miR-18a processing by hnRNP A1. Adv Exp Med Biol 700:28–35
29. Kawahara Y, Mieda-Sato A (2012) TDP-43 promotes microRNA biogenesis as a component of the Drosha and Dicer complexes. Proc Natl Acad Sci U S A 109(9):3347–3352. https://doi.org/10.1073/pnas.1112427109
30. Wu SL, Fu X, Huang J, Jia TT, Zong FY, Mu SR, Zhu H, Yan Y, Qiu S, Wu Q, Yan W, Peng Y, Chen J, Hui J (2015) Genome-wide analysis of YB-1-RNA interactions reveals a novel role of YB-1 in miRNA processing in glioblastoma multiforme. Nucleic Acids Res 43(17):8516–8528. https://doi.org/10.1093/nar/gkv779
31. Chang HM, Triboulet R, Thornton JE, Gregory RI (2013) A role for the Perlman syndrome exonuclease Dis3l2 in the Lin28-let-7 pathway. Nature 497(7448):244–248. https://doi.org/10.1038/nature12119
32. Faehnle CR, Walleshauser J, Joshua-Tor L (2014) Mechanism of Dis3l2 substrate recognition in the Lin28-let-7 pathway. Nature 514(7521):252–256. https://doi.org/10.1038/nature13553
33. Heo I, Joo C, Kim YK, Ha M, Yoon MJ, Cho J, Yeom KH, Han J, Kim VN (2009) TUT4 in concert with Lin28 suppresses microRNA biogenesis through pre-microRNA uridylation. Cell 138(4):696–708. S0092-8674(09)00964-7 [pii]. https://doi.org/10.1016/j.cell.2009.08.002
34. Hagan JP, Piskounova E, Gregory RI (2009) Lin28 recruits the TUTase Zcchc11 to inhibit let-7 maturation in mouse embryonic stem cells. Nat Struct Mol Biol 16(10):1021–1025. nsmb.1676 [pii]. https://doi.org/10.1038/nsmb.1676
35. Heo I, Joo C, Cho J, Ha M, Han J, Kim VN (2008) Lin28 mediates the terminal uridylation of let-7 precursor MicroRNA. Mol Cell 32(2):276–284. S1097-2765(08)00660-6 [pii]. https://doi.org/10.1016/j.molcel.2008.09.014
36. Newman MA, Thomson JM, Hammond SM (2008) Lin-28 interaction with the Let-7 precursor loop mediates regulated microRNA processing. RNA 14(8):1539–1549. rna.1155108 [pii]. https://doi.org/10.1261/rna.1155108
37. Treiber T, Treiber N, Plessmann U, Harlander S, Daiss JL, Eichner N, Lehmann G, Schall K, Urlaub H, Meister G (2017) A compendium of RNA-binding proteins that regulate MicroRNA biogenesis. Mol Cell 66(2):270–284 e213. https://doi.org/10.1016/j.molcel.2017.03.014

Chapter 10

Analysis of 3′ End Modifications in microRNAs by High-Throughput Sequencing

Madalena M. Reimão-Pinto, Angela M. Rodrigues-Viana, and Stefan L. Ameres

Abstract

MicroRNAs are ~22 nt small, non-coding RNAs that direct posttranscriptional silencing of gene expression to regulate animal development, physiology, and disease. An emerging mechanism that controls the biogenesis of microRNAs is the addition of non-templated nucleotides, predominantly uridine, to the 3′ end of precursor-microRNAs, in a process that is commonly referred to as tailing. Here, we describe methods that enable the systematic characterization of tailing events in mature microRNAs and their precursors. We report protocols for untargeted and targeted cDNA library preparation procedures, as exemplified in the context of the model organism *Drosophila melanogaster* and focusing on precursor-microRNAs. We also refer to a dedicated computational framework for the subsequent analysis of untemplated nucleotide additions in cDNA libraries. The described methods for the systematic characterization of posttranscriptional modifications in gene regulatory small RNAs and their precursors will be instrumental in clarifying regulatory concepts that control posttranscriptional gene silencing.

Key words Small RNA library preparation, Targeted pre-miRNA sequencing, 3′ end tailing analysis, Posttranscriptional RNA modifications

1 Introduction

Posttranscriptionally deposited chemical modifications can affect RNA fate, localization, and function [1]. Among the numerous RNA modifications that recently re-gained attention are 3′ terminal non-templated additions of ribonucleotides, also referred to as tailing. While canonical polyadenylation of mRNA was noted already in the early 1970s, it is only now becoming evident that multiple types of tailing, including non-canonical adenylation and uridylation, play important roles in the regulation of RNA fate and function [2]. The recent development of novel high-throughput methods now offers the opportunity to examine chemical modifications, including 3′ terminal nucleotide addition, even in low abundant RNA species at the genomic scale [3].

Ulf Andersson Vang Ørom (ed.), *miRNA Biogenesis: Methods and Protocols*, Methods in Molecular Biology, vol. 1823, https://doi.org/10.1007/978-1-4939-8624-8_10,

The addition of nucleotides to the 3′ end of RNA is catalyzed by ribonucleotidyltransferases, a protein family containing a characteristic catalytic domain with sequence homology to DNA polymerase β [2]. The genomes of both flies and mammals encode seven putative terminal nucleotidyltransferases (TNTases), but their enzymatic properties, substrate specificities, and biological functions remain poorly understood [4, 5]. Some TNTases exhibit a clear preference in terms of incorporated nucleotide identity, i.e., predominantly adenine for non-canonical poly(A) polymerases (ncPAPs) or uridine for terminal uridylyltransferases (TUTases), whereas others display more relaxed nucleotide specificities, catalyzing both uridylation and adenylation [6].

Depending on cellular compartment, RNA adenylation can have opposing function: For example, the addition of short A-tails triggers exosomal decay of aberrant transcripts in eukaryotic nuclei, whereas polyadenylation increases the stability and translatability of mRNA in the cytoplasm of eukaryotes [7, 8]. In contrast to adenylation, uridylation only recently emerged as a widespread cytoplasmic posttranscriptional mark associated with an expanding set of both coding and non-coding transcripts and frequently resulting in RNA destabilization [9].

Mounting evidence indicates that tailing of small RNAs and their precursors represents a major mechanism for the regulation of small RNA-mediated gene silencing [10]. Small RNA tailing was first described in *A. thaliana*, where methylation of small RNAs by HEN1 protects microRNAs (miRNAs) from uridylation and destabilization [11]. Since these initial observations, multiple studies have shown that not only mature miRNAs but also precursor miRNAs (pre-miRNAs) are subjected to posttranscriptional uridylation and adenylation in a variety of animals [5, 10–13]. Although it is evident that posttranscriptional tailing of small RNAs has profound implications on miRNA biogenesis and function, we still lack a comprehensive understanding of the functional relevance of many small RNA modifications. The development and/or adaptation of high-throughput sequencing technologies is a pre-requisite for investigating small RNA tailing events at the genomic scale. To this end, protocols for the generation of cDNA libraries from mature small RNAs are well established and have co-evolved with sequencing platforms [14–17]. Small RNA cloning approaches are typically based on (1) the physical separation of miRNAs from other more abundant RNA species (column based or by gel excision); (2) sequential ligation to oligonucleotide adaptors; (3) preparation of library cDNA; and (4) PCR amplification followed by high-throughput sequencing (Fig. 1). Because the average length of pre-miRNAs matches that of comparably very abundant species such as tRNAs and snoRNAs, selective isolation of pre-miRNA species by size exclusion is not possible. Thus, most studies addressing posttranscriptional pre-miRNA nucleotide

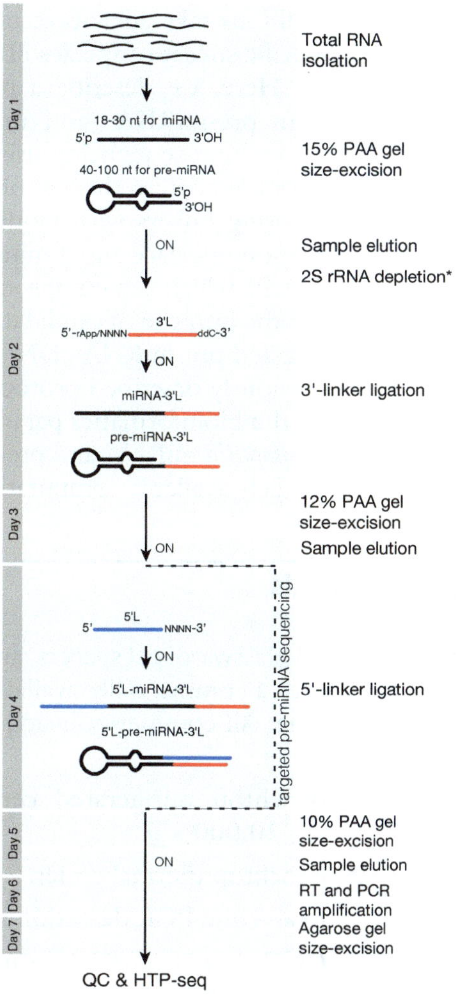

Fig. 1 Schematics of the small RNA library preparation protocol. After total RNA isolation and resolution on a 15% polyacrylamide gel, small RNAs are size-selected and eluted from the gel overnight (Subheading 3.1). Optional (indicated by star): If isolating mature miRNAs from *Drosophila* samples, an additional 2S ribosomal RNA depletion step is performed (Subheading 3.2). Small RNAs are ligated to a 3′ linker containing four random nucleotides at its 5′ end overnight (Subheading 3.3). 3′ linker-ligated small RNAs are resolved on a 12% polyacrylamide gel, size-selected and eluted from the gel overnight (Subheading 3.4). Small RNAs are subsequently ligated to a 5′ linker containing four random nucleotides at its 3′ end overnight (Subheading 3.5), followed by resolution on a 10% polyacrylamide gel, size-excision and elution overnight (Subheading 3.6). Small RNAs are then subjected to reverse transcription and PCR amplification followed by agarose gel size-excision (Subheadings 3.7 and 3.8). If performing the targeted pre-miRNA sequencing approach (Subheading 3.9), the 5′ linker ligation step is omitted. Samples are then ready for quality control and high-throughput sequencing (Subheading 3.10). The duration of the protocol steps is specified on the vertical gray bars. *PAA* polyacrylamide; *rRNA* ribosomal RNA; *3′L* 3′ linker; *5′L* 5′ linker; *QC* quality control; *HTP-seq* high-throughput sequencing

additions take advantage of targeted sequencing approaches for a specific subset of species [18–25].

Here, we describe a protocol for cDNA library preparation from pre-miRNA (and mature miRNAs) for subsequent high-throughput sequencing using the Illumina platform. This protocol does not require the use of radioactivity, minimizes ligation biases by using linkers with random nucleotide sequences at the RNA-ligation interface and diminishes the generation of "empty" adapter dimers, which can compromise library amplification yield [17, 26]. For the purpose of candidate pre-miRNA analysis, we also present a targeted pre-miRNA cDNA library preparation protocol, based on previously described protocols [24, 25]. Finally, we provide details on the bioinformatics parameters routinely used for the analysis of *Drosophila* miRNA and pre-miRNA tailing events, by means of the publicly available computational pipeline "Tailor" [27].

2 Materials and Reagents

2.1 Equipment

All glassware, gel spacers, and combs should be thoroughly cleaned with a commercially available detergent in distilled water ahead of time. All commercially acquired plasticware should be RNase-free grade.

Benchtop refrigerated centrifuge capable of speeds of up to 20,000 × *g*.

Benchtop thermal cycler.

Real-Time PCR detection system.

CCD-camera with UV-compatible emission filter.

UV-transilluminator.

Upright vertical gel electrophoresis system.

Horizontal gel electrophoresis system.

High-voltage electrophoresis power supply.

Heat block.

Spatula.

Aluminum plate.

Magnetic stand for 1.5 mL tubes.

Tube rotator for 1.5 and 2 mL tubes.

Sterile dounce homogenizer and motorized pestle mixer.

Safe-lock 1.5 mL and 2 mL tubes.

0.2 mL 8-tube PCR strips and ultraclear flat cap strips.

Filter tips (1–1000 μL).

Rain-X Original Glass Water Repellent and Interior Glass Anti-Fog.

1¼″ binder clips (four per gel).

Glass-plate set, 20 cm × 16 cm × 5 mm (length × width × thickness).

16-well comb and spacer set, 1.5 mm and 1 mm thick.

50 mL falcon tubes.

Volumetric pipettes (5, 10, and 25 mL).

Agarose gel tray 15 cm × 10 cm (length × width) and 15-well comb, 1 mm thick.

Surgical blades.

Microvolume Spectrophotometer.

2.2 Oligonucleotides and Gel Markers

RNA Marker.

50 bp DNA ladder.

Linkers and 18, 30, 40, and 60 mers oligonucleotides should be gel purified in advance (*see* **Note 1**). For a protocol, see [28]. For oligo sequences, see Table 1.

2.3 RNA Isolation

TRIzol® reagent or miRNA Isolation Kit.

Chloroform:isoamyl alcohol (24:1).

3 M NaOAc (pH 5.2).

Isopropanol.

Glycogen (20 mg/mL).

2.4 RNA Size Selection

All solutions should be prepared using MilliQ ultrapure water (prepared by purifying deionized water to attain 18.2 MΩ cm at 25 °C).

SequaGel UreaGel System.

10% APS (w/v).

TEMED.

0.5× Tris-Borate-EDTA (TBE) buffer: 45 mM Tris–Cl, 45 mM boric acid, 1 mM EDTA, pH 8.3.

Gel Loading Buffer II (Ambion).

SYBR® Gold Nucleic Acid Gel Stain (Life Technologies).

Elution buffer: 0.3 M NaCl, 0.1% SDS, filter sterilized with a 0.22 μm filter.

Ethanol.

2.5 2S rRNA Depletion

This step only applies to miRNA samples isolated from *Drosophila*.

My One streptavidin C1 beads (10 mg/mL).

20× SSC buffer: 3 M NaCl in 0.3 M sodium citrate, pH 7.0.

Table 1

"rApp" indicates adenylation and "ddC" indicates dideoxy; N denotes random nucleotide sequences or barcodes

Name	Type	Sequence (5′-3′)
2S rRNA-depletion oligo	DNA	Biotin-AGTCTTACAACCCTCAACCATATGTAGTCCAAGCAGCACT
3′ linker	DNA	rApp/NNNNAGATCGGAAGAGCACACGTCT/ddC/
5′ linker	RNA	ACACUCUUUCCCUACACGACGCUCUUCCGAUCUNNNN
RT primer	DNA	GTGACTGGAGTTCAGACGTGTGCTCTTCCGATCT
PCR forward TruSeq Univ. Adapter primer[a]	DNA	AATGATACGGCGACCACCGAGATCTACACTCTTTCCCTACACGACGCTCTTCCGATCT
PCR reverse TrueSeq Adapter Index primer[a]	DNA	CAAGCAGAAGACGGCATACGAGAT*NNNNNN*GTGACTGGAGTTCAGACGTGTGCTCTTCCGATCT
18 mer	RNA	GGUGGACAUCACUUACGC
30 mer	RNA	AGUAUGGGCAUUUCGCAGCCUACCGUGGUG
40 mer	RNA	AGCCUUCAGGAUUACAAGAUUCAAAGUGCGCUGCUGGUGC
60 mer	RNA	GGCUUGUCUGGCCUUUCACUACUCCUACGAGCACCAAGACAAGAUCAAGGCCAUCGUCCA
Pre-miRNA RT primer	DNA	ACTGGAGTTCAGACGTGTGCTC

2.6 RNA Linker Ligation

All buffers should be prepared using ultrapure water and subsequently filter sterilized with a 0.22 μm filter.

10× 3′ linker ligation buffer: 500 mM Tris pH 7.5, 100 mM $MgCl_2$, freshly added 100 mM DTT.

10× 5′ linker ligation buffer: 500 mM Tris pH 7.8, 100 mM $MgCl_2$, freshly added 100 mM DTT and 10 mM ATP.

T4 RNA Ligase truncated KQ (NEB).

T4 RNA ligase (Ambion).

50% PEG (8000).

2.7 Reverse Transcription and Library Amplification

SuperScript III Reverse Transcriptase kit (Invitrogen), containing 5× First-Strand Buffer and 0.1 M DTT.

KAPA HiFi HotStart ReadyMixPCR kit.

GoTaq Green Master Mix.

2.8 Gel Purification of Library cDNA

Certified Low range Ultra Agarose.

SYBR Safe DNA Gel Stain.

6× Orange Gel Loading Dye.

1× Tris-Acetate-EDTA (TAE) buffer: 40 mM Tris-acetate and 1 mM EDTA, pH 8.3.

Gel DNA Recovery Kit.

3 Methods

Unless otherwise specified, procedures are carried out at room temperature. For miRNA and pre-miRNA expression and tailing analysis, we perform libraries for each experimental condition in technical triplicates (each with a different barcode; see Subheading 3.8). We recommend being particularly careful when handling samples and taking measures against RNase contamination. We clean the working bench and pipettes with 70% ethanol before starting library preparation as a standard precaution procedure. Carefully follow all local hazardous waste disposal regulations. The protocol here described, from total RNA isolation to small RNA library PCR amplification, can be completed within 7 days (Fig. 1). When preparing small RNA libraries, we avoid pausing the protocol in between any of the steps; however, RNA samples can be precipitated at −20 °C for longer periods of time than the ones described in the protocol, or be stored at −80 °C in between steps when resuspended in water. We would like to point out that several library preparation kits are commercially available. For studies comparing methods, see [29, 30].

3.1 Small RNA Size Selection from Total RNA

1. Extract RNA with TRIzol® reagent as specified by manufacturer; work under a fume-hood.

   We routinely collect 1 × 10^7 Drosophila S2 cells yielding total RNA amounts in the range of 100–120 μg (*see* **Note 2**).

2. Before assembling the gel-casting apparatus, clean the surface of the glass plates that will be in contact with the acrylamide gel with 70% ethanol. Coat the rectangular plate with Glass Anti-Fog solution and the notched plate with water repellent solution. Assemble the glass plates by vertically positioning the spacers and clamp binder clips at all four corners.

3. Prepare a 1.5-mm thick 15% denaturing polyacrylamide gel using the UreaGel system (*see* **Note 3**). For each gel, prepare 55 mL of solution by mixing 33 mL of UreaGel Concentrate, 16.5 mL UreaGel Diluent, and 5.5 mL UreaGel Buffer. To initiate the polymerization reaction, add 330 μL 10% APS (w/v) and 33 μL TEMED and mix well. Immediately pour the acrylamide solution into the glass plates, insert the comb, and let the gel polymerase for ~1 h (*see* **Note 4**). After set, carefully remove the comb. Remove the two lower clamp binders, assemble the gel onto the gel rig, and clamp an aluminum plate to the glass to ensure even distribution of heat during the run. Add 0.5× TBE and rinse wells by flushing buffer with a pipette to displace accumulated urea. Pre-run gel for 45 min at 30 W or until the metal plates heat to 50–55 °C.

4. Prepare two RNA gel markers: one containing 0.5 μg RNA Century Plus marker and 10 pmol 40 mer and another containing 18 and 30 mer (10 pm each) (see Fig. 2a) diluted in 20 μL MilliQ water. Add 20 μL Gel Loading Buffer II and keep markers on ice. Prepare sample for loading by mixing 20 μg of total RNA with an equal volume of Gel Loading Buffer II (a 16-well, 1.5-mm thick comb has a loading capacity of approximately 60 μL total volume per well). Denature RNA samples and markers by incubating them on a 95 °C heat block for 5 min; spin down and keep samples on ice. Stop the gel pre-run, re-rinse the wells, and carefully load the gel markers and samples into the wells. To avoid sample cross-contamination during library preparation, we load a maximum of two samples per gel, at least two wells apart (Fig. 2a). Run the gel at 25 W for approximately 45 min, or until the bromophenol-blue band reaches about ¾ of the gel (*see* **Note 5**).

5. When the run is complete, prepare a staining solution (1:10,000 dilution of SYBR Gold Nucleic Acid Gel Stain in 25 mL 0.5× TBE). Remove the gel from the apparatus, and place it on a clean surface with the notched plate facing up. Insert a spatula in between the two plates to separate them; the acrylamide gel will stick to the bottom glass plate. Pour the staining solution onto the gel and incubate for 5 min (*see* **Note 6**). Pour away

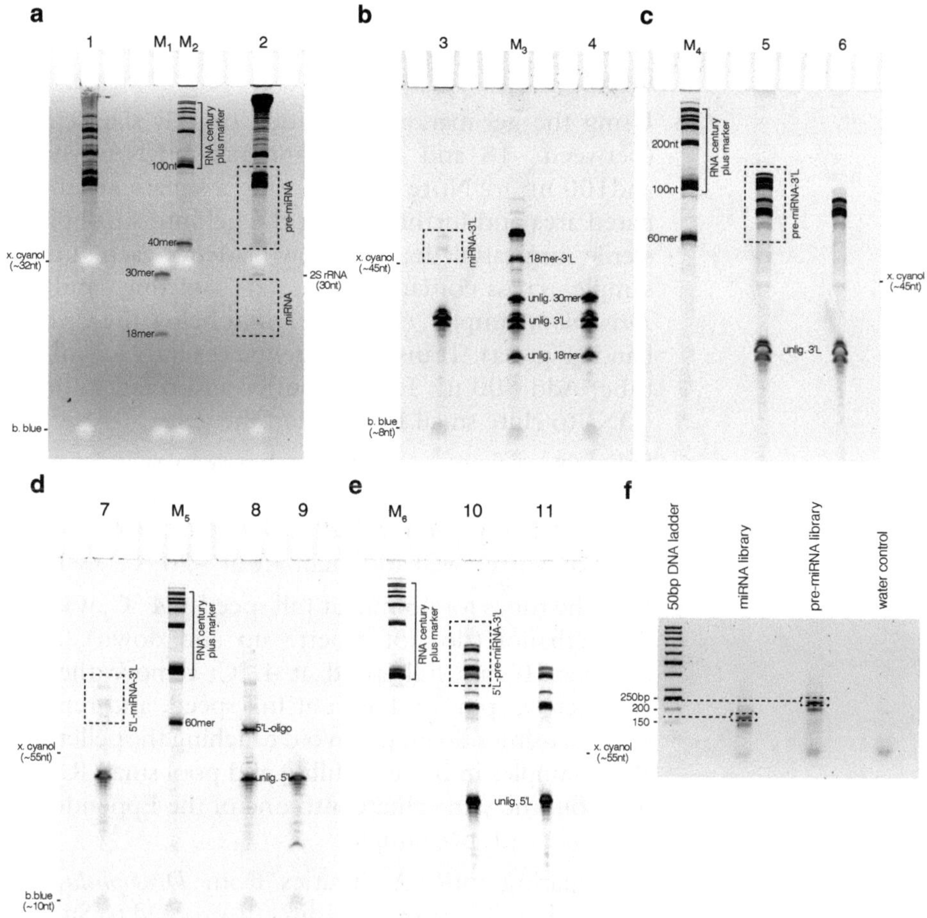

Fig. 2 Polyacrylamide and agarose gel excision steps. (**a**) Total RNA from mouse embryonic stem cells (1) and *Drosophila* S2 cells (2) was resolved on a 15% polyacrylamide gel. M_1 consists of 10 pmol 18 mer and 10 pmol 30 mer; M_2 consists of 0.5 μg RNA Century Plus marker and 10 pmol 40mer. Dashed lines represent the areas that were gel-excised. Note the presence of the 2S rRNA band that should not be gel-excised (2). (**b**) 3′ linker-ligated miRNA (3) was resolved on a 12% polyacrylamide gel and the area represented by the dashed lines was gel-excised. M_3 consists of 3′ linker-ligated 18 mer and 30 mer; a non-ligation control reaction was included (4). (**c**) 3′ linker-ligated pre-miRNA (5) was resolved on a 12% polyacrylamide gel and the area represented by the dashed lines was gel-excised; a non-ligated pre-miRNA control sample was included (6); M_4 consists of 0.5 μg RNA Century Plus marker and 5 pmol 60mer. (**d**) 5′ linker-ligated miRNA (7) was resolved on a 10% polyacrylamide gel and the area represented by the dashed lines was gel-excised. M_5 consists of 0.5 μg RNA Century Plus marker and 5 pmol 60mer. A 5′ linker control reaction consisting of the ligation of the 5′ linker to a 20 nt long 5′-phosphorylated-3′-inverted abasic oligonucleotide (8) and the corresponding non-ligation control (9) were included. (**e**) 5′ linker-ligated pre-miRNA (10) was resolved on a 10% polyacrylamide gel and the area represented by the dashed lines was gel-excised; a non-5′ linker-ligated, 3′ linker-ligated pre-miRNA control sample was included (11); M_6 consists of 0.5 μg RNA Century Plus marker. (**f**) PCR amplified miRNA and pre-miRNA samples were resolved on a 2% low-range agarose gel and gel-excised as indicated by the dashed lines. *3′L* 3′ linker; *5′L* 5′ linker; *unlig* unligated; *x. cyanol* xylene cyanol; *b. blue* bromophenol blue

excess staining solution and image the stained gel using a CCD-camera with a UV-compatible emission filter; record the image.

6. Using the gel markers as guides, quickly size select miRNAs (between ~18 and 30 nt) and pre-miRNAs (between ~40 and100 nt; *see* **Note 7**) by cutting a square around the designated area and further slicing the gel into small pieces using a sterile surgical blade; use a new blade for each sample to avoid sample cross-contamination. If isolating miRNAs from *Drosophila* samples, cut the gel right below the visible 2S rRNA band (Fig. 2a). Transfer the gel pieces into a 2 mL Eppendorf tube. Add 800 μL Elution buffer and rotate tubes overnight (ON) to elute small RNAs from the gel.
7. Split eluate from each sample between two 1.5 mL reaction tubes (i.e., approximately 400 μL each) (*see* **Note 8**). Precipitate small RNAs by adding 1 μL Glycogen and 2.5 volumes 100% ethanol; vortex well and incubate at −20 °C (*see* **Note 9**).
8. Spin the tubes for 45 min at full speed at 4 °C, wash pellet with 80% ethanol (do not pipette up and down), and spin for another 10 min full speed at 4 °C; remove the supernatant completely, spin for 1 min at full speed, and remove residual ethanol using a 10 μL tip (avoid touching the pellet). Resuspend RNA samples in 5.5 μL MilliQ and pool small RNAs originating from the same eluate into one of the Eppendorf tubes to a total of 11 μL per sample.
9. If preparing miRNA libraries from *Drosophila*, proceed to Subheading 3.2; otherwise directly proceed to Subheading 3.3 (3′ linker ligation).

3.2 *Drosophila* 2S Ribosomal RNA Depletion

This step only applies to mature (i.e., 18–30 nt size selected) miRNA samples isolated from *Drosophila*, whose genome encodes for 30 nt long 2S ribosomal RNA (2S rRNA) [31]. The gel size selection is not sufficient to completely exclude this highly abundant rRNA, and as such 2S rRNA depletion is achieved by annealing it to biotinylated antisense oligos, followed by streptavidin-coated beads purification (protocol adapted from [32]).

1. Dilute 20× SSC buffer to 0.5× SSC in MilliQ in advance; keep on ice. Spin the tubes for 45 min at full speed at 4 °C, wash pellet with 80% ethanol (do not pipette up and down), and spin for another 10 min full speed at 4 °C; remove the supernatant completely, spin for 1 min at full speed, and remove residual ethanol using a 10 μL tip (avoid touching the pellet). Resuspend small RNAs in 5 μL 0.5× SSC and pool small RNAs originating from the same sample (see Subheading 3.1, **step 7**) into one of the Eppendorf tubes to a total of 10 μL per sample. Keep samples on ice.

2. Add My One streptavidin C1 beads (160 μL beads per sample) to a 1.5 mL tube and collect beads using a magnetic stand. Remove the supernatant and wash beads in 1 mL pre-chilled 0.5× SSC by gently pipetting up and down; repeat for a total of two washes. Resuspend beads in 160 μL 0.5 μM *2S rRNA-depletion oligo* in 0.5× SSC; incubate for 30 min on ice.
3. Capture beads using the magnetic stand, discard supernatant, and wash beads three times with 1 mL 0.5 × SSC. Resuspend beads in 70 μL 0.5× SSC and incubate for 5 min at 65 °C. In parallel, denature the RNA by heating the sample for 5 min at 80 °C. Immediately after, add the miRNA sample directly to the beads and mix well by flicking the tube. Incubate for 1 h at 50 °C.
4. Capture beads and transfer the 2S rRNA-depleted supernatant to a clean tube, avoiding any bead carryover. Precipitate the purified miRNA by adding 1 μL Glycogen, 1/10 volume of 3 M NaOAc (pH 5.2), and 2.5 volumes 100% ethanol; vortex well and incubate at −20 °C (*see* **Note 9**). Spin and wash RNA as before (Subheading 3.1, **step 8**) and resuspend in 11 μL MilliQ water.

3.3 3′ Linker Ligation

Linker ligation steps are a main source for biases in small RNA library read representation. We use linker oligos containing four random nucleotides at the RNA-ligation interface, thus minimizing ligation biases [33–35]. Ligation reactions are targeted to small RNAs containing a 5′ phosphate and a 3′ hydroxyl termini. For small RNA 3′ linker ligation, we use a truncated version of the T4 RNA ligase 2 (T4 Rnl2) enzyme that contains two point mutations, R55K and K227Q at its active sites, which prevent the formation of ligation side products such as input RNA concatemers [15, 36]. Pre-adenylation of the 5′-end of the 3′ linker renders it a substrate for T4 Rnl2, thus addition of ATP to the ligation reaction is not required (see Subheading 2.6). Furthermore, the 3′ linker contains a dideoxy terminator at the 3′ end to prevent adaptor circularization. As a control for 3′ linker ligation reaction efficacy and for subsequent gel size selection (Fig. 2b), the 18 and 30 mers are ligated to the 3′ linker in a parallel reaction. When performing libraries only for pre-miRNA samples, we divide the sample into two—3′ linker-ligated pre-miRNA sample and non-ligated control sample—to control for 3′ linker ligation efficacy and for gel size selection of 3′ linker-ligated pre-miRNA sample (Fig. 2c; *see* **Note 10**).

1. Spin down and wash RNA as in Subheading 3.1, **step 8**. Resuspend pellet in a total of 11 μL MilliQ water. Keep samples on ice.
2. Prepare the 3′ linker ligation reaction (*see* **Note 11**):

Prepare enough 3′ linker ligation master mix for additional ligation reactions: for 18 and 30 mer size markers, and/or for 40 mer size marker (*see* **Note 10**). For preparing the ligation reactions for the size markers, replace the RNA sample with 25 pmol of 18 and 30 mer each, or 25 pmol 40 mer. Mix thoroughly by pipetting and incubate ON at 4 °C.

50% PEG (8000)	5 μL
3′ linker (50 μM)	1 μL
10× 3′ linker ligation buffer	2 μL
T4 Rnl2tr R55K, K227Q (200 U/μL)	1 μL
RNA sample	11 μL
Total	20 μL

3.4 Gel Purification of 3′ Linker-Ligated Small RNAs

To avoid the downstream ligation of the 3′ linker directly to the 5′ linker to produce "empty" adaptor dimers, we remove unligated linkers by denaturing polyacrylamide gel excision after each linker ligation step.

1. Prepare the gel-casting material as in Subheading 3.1, **step 2**. Prepare a 1-mm thick 12% denaturing polyacrylamide gel using the UreaGel system. For each gel, prepare 40 mL of solution by mixing 19.2 mL of UreaGel Concentrate, 16.8 mL UreaGel Diluent, and 4 mL UreaGel Buffer. To initiate the polymerization reaction, add 240 μL 10% APS (w/v) and 24 μL TEMED and mix well. Immediately cast the gel and proceed as previously described (Subheading 3.1, **step 3**).
2. Add 20 μL Gel Loading Buffer II to the 3′ linker-ligated samples and mix well; if cloning pre-miRNAs, prepare at this step an RNA ladder by mixing 0.5 μg RNA Century Plus marker and 5 pmol 60 mer in 20 μL MilliQ water and add 20 μL Gel Loading Buffer II (*see* **Note 12**). Incubate samples on a 95 °C heat block for 5 min, spin down and keep them on ice. Stop the pre-run, re-rinse the wells, and carefully load the gel markers and samples into the wells. To avoid sample cross-contamination during library preparation, we load a maximum of two samples per gel, at least two wells apart (Fig. 2b, c). Run the gel at 25 W for approximately 45 min, or until the xylene cyanol band reaches about ½ of the gel (*see* **Note 13**).
3. Mature miRNA samples ligated to the 3′ linker should be ~46–49 nt, and 3′ linker-ligated pre-miRNAs should be ~65–125 nt. For 3′ linker-ligated miRNA size selection, cut the gel between the 3′ linker-ligated 18 and 30 mer size markers (see Subheading 3.1, **step 5** for details; Fig. 2b). For 3′ linker-ligated pre-miRNA size selection, cut between the 60 mer (or 3′ linker-ligated 40 mer; *see* **Note 12**) and ~150 nt, as referred to by the RNA Century Plus marker (Fig. 2c; *see* **Note 14**). For eluting

and precipitating small RNAs, proceed as in Subheading 3.1, **steps 6** and 7, respectively. For the targeted pre-miRNA sequencing approach, proceed directly to the reverse transcription step (Subheading 3.7).

3.5 5′ Linker Ligation

In this step, ligation of the 5′ linker to the 5′ phosphate of 3′ linker-ligated small RNAs is achieved by T4 RNA ligase 1. Successful ligation requires the addition of ATP (Subheading 2.6) which will not result in input RNA or 5′ linker circularization side products, as the 3′ end of the 3′ linker is rendered inactive by the dideoxy termini and the 5′ linker is not phosphorylated at its 5′end [36].

1. Spin down and wash 3′ linker-ligated RNA as in Subheading 3.1, **step 8**. Resuspend RNA samples in 5.5 μL MilliQ and pool small RNAs originating from the same eluate (see Subheading 3.1, **step** 7) into one of the Eppendorf tubes to a total of 11 μL per sample. Keep samples on ice.

   When performing libraries only for pre-miRNA samples, we divide the sample into two—5′ linker-ligated pre-miRNA sample and non-ligated control sample—to control for 5′ linker ligation efficacy and for gel size selection of 5′ linker-ligated pre-miRNA sample (Fig. 2e). Alternatively, a 5′ phosphorylated DNA or RNA oligonucleotide can be used to control for 5′ linker ligation efficacy (Fig. 2d; *see* **Note 15**).
2. Prepare the 5′ linker ligation reaction (*see* **Note 11**):

   Mix thoroughly by pipetting and incubate ON at 4 °C.

50% PEG (8000)	5 μL
5′ linker (50 μM)	1 μL
10× 5′ linker ligation buffer	2 μL
T4 RNA ligase (5 U/μL)	1 μL
RNA sample	11 μL
Total	20 μL

3.6 Gel Purification of 5′ Linker-Ligated Small RNAs

1. Prepare the gel-casting material as in Subheading 3.1, **step 2**. Prepare a 1-mm thick 10% denaturing polyacrylamide gel using the UreaGel system. For each gel, prepare 40 mL of solution by mixing 16 mL of UreaGel Concentrate, 20 mL UreaGel Diluent, and 4 mL UreaGel Buffer. To initiate the polymerization reaction, add 240 μL 10% APS (w/v) and 24 μL TEMED and mix well. Immediately cast the gel and proceed as previously described (Subheading 3.1, **step 3**).
2. Add 20 μL Gel Loading Buffer II to the 5′ linker-ligated samples and mix well; for miRNA size selection, prepare a size marker by mixing 0.5 μg RNA Century Plus marker with 5 pmol 60 mer in 20 μL MilliQ and 20 μL Gel Loading Buffer

II. For pre-miRNA size selection, add 0.5 μg RNA Century Plus marker to 20 μL MilliQ and 20 μL Gel Loading Buffer II. Mature miRNA samples ligated to the 5′ linker should be ~83–86 nt in length, and 5′ linker-ligated pre-miRNAs should be ~102–162 nt. For 5′ linker-ligated miRNA size selection, cut the gel between the 60 mer and 100 nt size markers (see Subheading 3.1, **step 5** for details; Fig. 2d). For 5′ linker ligated pre-miRNA size selection, cut between the 100 and 200 nt band of the RNA Century Plus marker (Fig. 2e). For eluting and precipitating small RNAs, proceed as in Subheading 3.1, **steps 6** and 7, respectively.

3.7 Small RNA Library Reverse Transcription

1. Spin down and wash linker-ligated RNA as in Subheading 3.1, **step 8**. Resuspend RNA samples in a total of 11 μL MilliQ water. Keep samples on ice.
2. Prepare the reverse transcription (RT) reaction mix on ice:

5× FS Buffer	4 μL
10 mM dNTPs	1 μL
0.1 M DTT	2 μL
SuperScript III RT (200 U/μL)	1 μL
Total	8 μL

3. Transfer the RNA sample to a 0.2 mL PCR strip. Add 1 μL of 50 μM *RT primer* to the RNA sample and mix well by pipetting. Incubate on a thermocycler for 2 min at 72 °C to denature the RNA. Immediately transfer to ice and incubate for 2 min to prevent the RNA from folding into secondary structures. Spin down sample quickly and keep on ice.
4. Add 8 μL of the RT reaction mix to the sample (for a final volume of 20 μL); mix well by pipetting. Transfer tube to a thermocycler and incubate for 60 min at 50 °C, followed by heat inactivation for 15 min at 70 °C. The cDNA can be directly used for PCR amplification or stored at −20 °C. For targeted pre-miRNA PCR amplification, proceed to Subheading 3.9.

3.8 Small RNA Library PCR Amplification

Library amplification is performed using the KAPA HiFi HotStart ReadyMixPCR Kit, which supplies a KAPA HiFi HotStart DNA Polymerase 2× ready mix and four fluorescent standards to define a window for optimal amplification. During library amplification, we systematically achieve optimal library amplification signal at around 6–8 or 10–12 cycles, for pre-miRNA and miRNA samples, respectively. We allow amplification to proceed until fluorescence intensity reaches a level in between standard #3 and #4 provided with the kit (~3400–3600 RFU), before plateauing of the amplification curve. Sample barcoding (PCR rev primer TruSeq) is introduced at this step, which precludes bias in miRNA expression profiles [26, 36, 37], and allows for sample multiplexing at the

HTP-Seq stage. The resulting small RNA libraries are purified by agarose gel size selection and subjected to quality control.

1. Set up the PCR library amplification reaction (*see* **Note 16**):

   Keep the master mix and the samples on ice. Transfer the RNA sample to a 0.2 mL PCR tube/strip compatible with your qPCR thermal cycler equipment, add the PCR amplification reaction mix, and mix well by pipetting. Spin down tube/strip (*see* **Note 17**).

2× KAPA HiFi HotStart	25 μL
PCR fwd primer TruSeq (25 μM)	1 μL
PCR rev primer TruSeq (25 μM)	1 μL
cDNA library	10 μL
MilliQ water	13 μL
Total	50 μL

2. Perform library amplification according to the following cycling conditions:

   When the library reaches the optimal amplification signal, pause the cycling reaction in the machine and allow for the sample to remain at 72°C for an extra 20" (for a total of ~1 min final extension step). Remove the sample and keep on ice if continuing with library purification, or store at −20 °C. If amplifying several samples simultaneously, continue the cycling reaction and go through this process until all samples are amplified according to individual optimal cycle number.

98°C	45 s	
98°C	15 s	
65 °C	30 s	X cycles (see text)
72°C	20 s (followed by SYBRGreen measurement)	
72°C	15 s	

3. Prepare a 2% low-range agarose gel containing SYBR™ Safe DNA Gel Stain (1:10,000 dilution). Ensure that the gel has a loading capacity of 60 μL sample per well. Add 10 μL of 6× Orange Gel Loading Dye to the sample (*see* **Note 18**); if purifying several libraries, leave one lane empty in between samples. Load the 50 bp DNA ladder and sample on the gel. Run the gel at 80 V for approximately 1 h 30 min until the Orange Gel Dye reaches ¾ of the gel (*see* **Note 19**). Image the gel and record the image; miRNA and pre-miRNA libraries run at approximately 170 nt and 230 nt in length, respectively (Fig. 2e). Cut out the library from the gel carefully excluding adaptor dimers and transfer the sample to a 1.5 mL Eppendorf tube using a sterile surgical blade; use a new blade for each sample to avoid sample cross-contamination. Elute the sample with the

Zymoclean™ Gel DNA Recovery Kit and resuspend small RNA libraries in 15 μL MilliQ water and store at −20 °C. Libraries are now ready for quality control (Subheading 3.10).

3.9 Targeted Pre-miRNA PCR Amplification

Global analysis of pre-miRNAs using a non-targeted sequencing approach as described above suffers from the fact that only 0.1% of all sequenced RNA species correspond to pre-miRNAs [5]. To be able to perform high-resolution annotation of tailing events for individual pre-miRNAs we use a targeted pre-miRNA sequencing approach, based on 3′ linker ligation of size selected total RNA followed by sequence-specific second strand synthesis [24, 25]. Note that this approach requires prior knowledge on pre-miRNA sequences (e.g., available via mirBase [38]).

Library amplification is performed using the GoTaq Green Master Mix and consists of two amplification steps. First, individual pre-miRNAs of interest are amplified using a pre-miRNA-specific primer (±25–30 nt) that binds to the loop and 3p arm (or alternatively the 5p arm) of the miRNA (see Table 2 for example primers). By only amplifying the 3p arm of each miRNA, the sequencing length required for characterization of miRNA 3′ end additions can be reduced to single-read 50 sequencing mode. If primers are designed to bind to the 5p arm of a pre-miRNA, single-read 100 sequencing mode is required to analyze 3′ end modifications (see Table 2).

1. Set up the PCR library amplification reaction:

2× GoTaq® Green Master Mix	12.5 μL
Forward pre-miRNA-specific primer (10 μM)	0.5 μL
Pre-miRNA RT primer (10 μM)	0.5 μL
cDNA RT reaction	1 μL
H_2O	10.5 μL
Total	25 μL

Prepare all components in PCR tubes on ice, mix well by pipetting up and down, spin down, and perform library amplification according to the following cycling conditions:

95 °C	2 min	
95 °C	30 s	
X °C[a]	30 s	10 cycles
72 °C	15 s	
72 °C	5 min	
4 °C	Hold	

[a]Annealing temperature varies for each pre-miRNA-specific forward primer and needs to be adjusted. Annealing conditions can be optimized by performing PCR gradients starting with temperature approximately 5 °C below the calculated melting temperature and increase in increments of 1 °C

Table 2

Primer sequences for the targeted amplification of pre-miRNAs after 3′ adapter ligation and reverse transcription

Pre-miRNA	Fwd pre-miRNA-specific sequence (DNA, 5′-3′)
dme-*bantam*	CACGACGCTCTTCCGATCTTTCATACAAGTGAGATCATTTTGAAAGC
dme-mir-1003	CACGACGCTCTTCCGATCTGGCGGTCCTCTCACATTTACATATTC
dme-mir-1006	CACGACGCTCTTCCGATCTTGGTACAATTTAAATTCGATTTCTTATTCA
dme-mir-1007	CACGACGCTCTTCCGATCTGGACTCTTGATAAGCTCAATTAACTGTT
dme-mir-1008	CACGACGCTCTTCCGATCTTGGCAAGTCACAGCTTTTTGTGTTT
dme-mir-1013	CACGACGCTCTTCCGATCTGGCCGTTAATAAAAGTATGCCGAACT
dme-mir-12	CACGACGCTCTTCCGATCTTCCAACAACCAGTACTTATGTCATACT
dme-mir-275	CACGACGCTCTTCCGATCTATACAGTCAGGTACCTGAAGTAGC
dme-mir-276a	CACGACGCTCTTCCGATCTAACTCGTAGGAACTTCATACCGTG
dme-mir-282	CACGACGCTCTTCCGATCTCCGATCAGACATAGCCTATAAGAGGT
dme-mir-283	CACGACGCTCTTCCGATCTTGAAACACTCGGAATTTCAGTTGGTAT
dme-mir-304	CACGACGCTCTTCCGATCTACGCACTCACTTTGCAATTGGAGAT
dme-mir-305	CACGACGCTCTTCCGATCTGTAACCCGGCACATGTTGAAGTAC
dme-mir-306	CACGACGCTCTTCCGATCTATTTTGGGGGTCACTCTGTGCCT
dme-mir-308	CACGACGCTCTTCCGATCTTCGTTTTGCAATCCAAATCACAGGATT
dme-mir-317	CACGACGCTCTTCCGATCTGCAAGTGAACACAGCTGGTGGTA
dme-mir-33	CACGACGCTCTTCCGATCT CTTCAGG CAATACAACTTCAGTGCA
dme-mir-34	CACGACGCTCTTCCGATCT ACAATTCA CAGCCACTATCTTCACTG
dme-mir-8	CACGACGCTCTTCCGATCT TTTATAACTC TAATACTGTCAGGTAAAG
dme-mir-988	CACGACGCTCTTCCGATCT GATCAT CCCCTTGTTGCAAACCTC

Underlined sequence indicates pre-miRNA-specific binding sequence; miRNA sequence is highlighted in blue. MicroRNA-specific sequence is preceded by 5′ adaptor sequence

2. Set up the second PCR library amplification reaction (*see* **Note 16**):

2× GoTaq® Green Master Mix	25 μL
PCR fwd primer TruSeq (10 μM)	1 μL
PCR rev primer TruSeq (10 μM)	1 μL
PCR 1 reaction	5 μL
H_2O	18 μL
Total	50 μL

Prepare all components in PCR tubes on ice, mix well by pipetting up and down, spin down, and perform library amplification according to the following cycling conditions:

95 °C	2 min	
95 °C	30 s	
55 °C	30 s	20 cycles[a]
72 °C	20 s	
72 °C	5 min	
4 °C	Hold	

[a]Note that amplification cycle-number requires optimization and optimal cycle numbers may vary depending on the abundance of the respective pre-miRNA

3. The resulting small RNA libraries are purified from agarose gels as described in Subheading 3.8, **step 3**. Purified PCR products are eluted in 15 μL MilliQ water.

3.10 Small RNA Library Quality Control and HTP-Seq

Prior to library HTP-seq, we control for the concentration and length profile of the small RNA libraries. We routinely use the PicoGreen dsDNA Quantitation Reagent (Turner BioSystems) for accurate measurement of library concentration and assess library size distribution using the High Sensitivity small DNA Fragment Analysis kit (Advanced Analytical; *see* **Note 20**). MicroRNA libraries show a sharp peak at 160–170 nt in length and pre-miRNA libraries at 210–230 nt. The concentration of each library is then adjusted to 5 nM based on these length and concentration analyses. Subsequently, we quantify diluted samples by qPCR using the KAPA Library Quantification Kit (KAPABiosystems) for Illumina using cycling conditions optimized for the iCycler Biorad machine. Library concentration is re-adjusted to 5 nM according to the final qPCR step and libraries are pooled together to an equimolar ratio in a concentration and amount required for the selected sequencing machine. For miRNA sequencing, we perform single-read 50 bp sequencing and for pre-miRNA libraries single-read 100 bp sequencing and aim for a minimum number of ten million and 70 million reads per library, respectively. For the targeted pre-miRNA approach, we perform single-read 50 bp and ten million reads suffice for tailing analysis. Libraries using the primer sets described in this protocol are compatible with Illumina platforms (here, we use a HiSeq2500 machine).

3.11 Bioinformatic Analysis

For miRNA and pre-miRNA tailing analysis, we use the publicly available computational framework Tailor (source code and executable binaries available at https://github.com/jhhung/Tailor) [27]. Tailor performs a sequence-based alignment analysis that returns the length and identity of non-templated additions to the

3′ end of transcripts. Alignments are assigned to different genomic features, meaning that tail information can be acquired not only for miRNAs, but for other non-coding and coding transcripts. The framework operates by interrogating the presence of tails at the 3′ end of a miRNA transcript by finding the longest common prefix between the sequenced read and the reference sequence for that miRNA. Tailor analyzes the reverse-complement of the query read starting from the 5′ end, and follows a set of rules to retain or discard a read. If it encounters at least one mismatch within a set number of nucleotides (5′-to-3′ direction) of the read, it assigns it as a genomic encoded sequence mismatch on the basis of possible editing events in the miRNA seed region. If the framework finds a second mismatch within the seed region, it discards the read. For cases where the mismatch occurs at the first nucleotide outside of the seed region and all following nucleotides match the genome, then it assigns it as a sequencing error; however, if the first mismatch outside of the seed region corresponds to the last nucleotide of the read (most 3′ end position), Tailor assigns it as a non-templated addition event. When the tail is longer than one nucleotide, Tailor further scans the query read to ensure that the tail consists of multiple non-templated events and provides information of the position and identity of each of the nucleotides comprising the tail. Note that Tailor cannot identify tailing events for miRNAs whose genomic encoded sequence matches that of the tail sequence. To further counteract internal mismatches, we apply an additional filtering step based on the ratio between tail length to sequence length; if this ratio is higher than 0.12 and 0.05, for miRNA and pre-miRNA sequencing, respectively, the read is discarded. Similar strategies can be applied to analyze targeted pre-miRNA libraries. For more details on the analysis, please refer to [5, 27].

3.12 Further Consideration and Interpretation of Results

While size selection of 18–30 nt long RNAs typically results in the majority of reads corresponding to bona fide small RNAs, predominantly mapping to miRNA, unbiased cloning for the characterization of pre-miRNAs suffers from the fact that abundant cellular non-coding RNA overlap in size with pre-miRNAs. Hence, the majority of reads in cDNA libraries generated from 40 to 100 nt long RNA map to tRNA, snoRNA, rRNA, and other non-coding RNA species, and less than 0.1% of all reads tend to map to pre-miRNAs [5]. Therefore, sequencing at sufficient depth is required for the analysis of pre-miRNA modifications. For example, ~70 mio reads in cDNA libraries generated from 40–100 nt RNAs derived from *Drosophila* S2 cells result in ~70,000 analyzable pre-miRNA mapping reads. This is sufficient to detect pre-miRNAs for the majority of all abundantly detected mature miRNAs, hence providing a comprehensive view on the pre-miRNome [5]. At the level of individual pre-miRNA species, a sequencing depth of ~24 mio reads is sufficient to clarify the posttranscriptional modification status of >50 pre-miRNAs (Fig. 3a) [5]. For example, in *Drosophila* S2 cells, signatures of 3′ tailing predominantly consist of uridylation and adenylation events (Fig. 3b). Strongest tailing is typically observed for pre-miRNA hairpins generated via the splicing and debranching, referred to as mirtrons (Fig. 3a, b). This reflects the specific targeting of mirtron hairpins by the TUTase Tailor, which prevents Dicer-processing and induces RNA decay via the uridylation-triggered 3′-to-5′ exoribonuclease Dis3l2, serving as a barrier for the de novo creation of miRNAs in Drosophila [5, 13, 39].

While targeted pre-miRNA libraries suffer from the fact that sequence information on expressed miRNAs is required for primer design and that resulting libraries do not readily provide insight into relative abundances, thousands of reads per pre-miRNA are sufficient to assess 3′ modification status. In fact, comparison of cDNA libraries generated from Drosophila S2 cells through the described untargeted and targeted approaches provides comparable results: Both the extent of tailing and the relative nucleotide identity for 20 pre-miRNAs (5 mirtrons and 15 canonical miRNAs) is similar in the targeted and the non-targeted approach (Fig. 3c, d). Hence, the initial characterization of pre-miRNA repertoires in a given cell type using the untargeted pre-miRNA library preparation protocol can readily be supplemented with a more detailed analysis of selected pre-miRNAs using the targeted approach in order to clarify the tailing status of miRNA precursors.

4 Notes

1. Custom RNA oligos can be ordered from commercial vendors or synthesized by in vitro transcription using e.g., T7 RNA polymerase; for a protocol, see ref. 40. For ease of use, after gel purification we prepare aliquots of each oligo at 10 μM concentration. RNA oligonucleotides should be placed at −80 °C for long-term storage.
2. Structured miRNAs (and thus potentially also pre-miRNAs) with low GC content have been shown to be selectively lost during total RNA isolation using this method, in particular when starting with a low cell number (see retraction note on [41]). As such, for low input material or for quantitative analysis of pre-miRNA abundance, we recommend using the *mir*Vana™ miRNA Isolation Kit according to manufacturer specifications

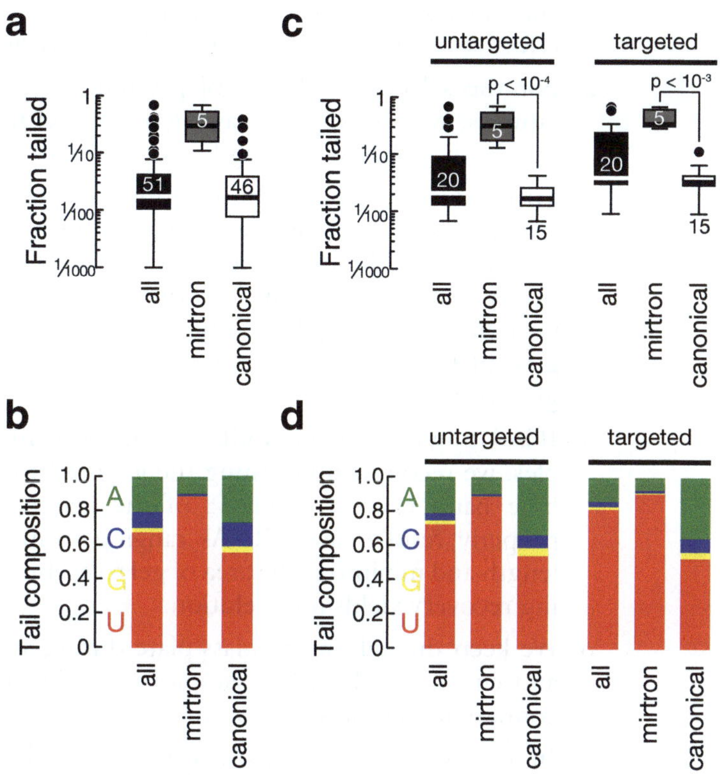

Fig. 3 Tailing analyses of pre-miRNA in cDNA libraries generated from size-selected total RNA using untargeted or targeted approaches. (**a**) and (**b**) Tailing analysis of 51 abundantly detected pre-miRNAs (>100 ppm) in non-targeted cDNA libraries generated from 40 to 100 nt size selected total RNA input derived from *Drosophila* S2 cells. (**a**) Boxplot shows the fraction of pre-miRNA reads that contain non-genome-matching 3′ end nucleotide additions for 51 abundantly detected pre-miRNAs (left), as well as five pre-miRNAs generated via the mirtron pathway (center) and 46 pre-miRNAs generated via the canonical miRNA pathway (right). (**b**) Nucleotide composition of non-genome-matching 3′ end additions for 51 abundantly detected pre-miRNAs (left), as well as five pre-miRNAs generated via the mirtron pathway (center) and 46 pre-miRNAs generated via the canonical miRNA pathway (right). (**c**) and (**d**) Comparison of tailing analysis of 20 abundantly detected pre-miRNAs in non-targeted (left) and targeted (right) cDNA libraries generated from 40 to 100 nt size selected total RNA input derived from *Drosophila* S2 cells. Analysis of non-targeted cDNA libraries was performed as previously described [5]. For targeted libraries, library reads were recovered by adapter clipping and removal of random 4mers on the 3′ end as previously described [5], followed by the identification of primer sequences in each read, allowing 10% mismatches, and correcting for potential mismatches in the primer sequence. To further counteract internal mismatches, aligned reads were filtered out if their ratios of tail length to sequence length were 0.07 or higher. (**c**) Boxplot shows the fraction of pre-miRNA reads that contain non-genome-matching 3′ end nucleotide additions for 20 selected pre-miRNAs (left), consisting of five pre-miRNAs generated via the mirtron pathway (center) and 15 pre-miRNAs generated via the canonical miRNA pathway (right). (**d**) Nucleotide composition of non-genome-matching 3′ end additions for 20 selected pre-miRNAs (left), consisting of five pre-miRNAs generated via the mirtron pathway (center) and 15 pre-miRNAs generated via the canonical miRNA pathway (right)

and include inert carrier supplement (i.e., Glycogen) for effective recovery of nucleic acids during precipitation.

3. For the selective isolation of pre-miRNAs from total RNA samples, we recommend casting a 10% gel for better resolution of small RNAs ranging from 40 to 100 nt in length.
4. To avoid introducing air bubbles into the gel when casting, we lay the assembled glass plates onto a tip box, notched plate facing up and slightly tilt it sideways. Use a 25 mL pipette to slowly but continuously introduce the acrylamide solution into the glass plates; the solution will migrate evenly across the glass plates and stop when it reaches the open bottom of the glass plates.
5. If isolating only mature miRNA samples from total RNA samples, we recommend running the gel until the bromophenol-blue band reaches about ½ of the gel; this is enough to properly resolve small RNAs ranging from 18 to 30 nt in length and minimizes the area of gel to excise, thereby improving recovery yields upon elution.
6. We keep the gel on the glass plate during staining, imaging, and cutting of the RNA samples to avoid ripping it and/or contaminating the samples.
7. In rare cases, pre-miRNAs can extend the size cutoff of 100 nts. We recommend consulting miRbase annotations for predicted pre-miRNA length [38].
8. The presence of acrylamide pieces can inhibit subsequent enzymatic reactions. To avoid carryover of small gel pieces, one can use gel loading tips to transfer the eluate. Alternatively, the eluate can be filtered using 0.22 μm pore size centrifugal units (e.g., Ultrafree-MC, Millipore).
9. For maximizing recovery of small RNAs, we typically precipitate samples ON; we recommend a minimum of 3 h of precipitation for miRNA samples and ON for pre-miRNA samples. We usually achieve 70–80% sample recovery after gel elution and precipitation under these conditions.
10. As an alternative to splitting the pre-miRNA sample into two (ligated and non-ligated control), the 40 mer can be ligated to the 3′ linker to control for ligation reaction efficacy.
11. PEG is a viscous solution that is cumbersome to mix. We usually set it to thaw at room temperature about 15 min prior to preparing the ligation reaction; we recommend preparing a master mix at least 10% in excess and to mix the ligation reaction by pipetting up and down slowly to avoid introducing air bubbles.
12. If performing a 40 mer-3′ linker ligation reaction control, add 20 μL Gel Loading Buffer II and 0.5 μg RNA Century Plus marker to the sample.

13. If only gel size selecting pre-miRNA samples, allow the gel to run for approximately 60 min or until the bromophenol-blue band reaches the bottom of the gel to allow for better resolution of 60–150 nt long small RNAs.
14. Note that it is not possible with this size selection step to physically separate successfully 3′ linker-ligated pre-miRNAs (~65–125 nt) from unligated pre-miRNAs (~40–100 nt). Carried-over unligated pre-miRNAs will likely be circularized when incubated with T4 RNA ligase and excluded from the library.
15. Oligonucleotide 3′ modifications such as inverted abasic (iB) end modification protect an oligo from concatemerization during 5′ linker ligation; to control 5′ adapter ligation, we commonly use the RNA oligo ACACGACGCUmCmUmUmCmCmGmAmUmCmUiB; m depicts 2′O-methylation modification.
16. If amplifying several libraries simultaneously, prepare a common master mix without the PCR rev primer TruSeq, and add a unique barcoding primer directly to each sample.
17. If amplifying multiple libraries using PCR strips, libraries might vary in their optimal number of PCR amplification cycles. As such, at this step we cut the plastic connecting each tube of the PCR strip to allow stopping the amplification reaction (i.e., removing the tube from the thermocycler) for each sample independently. We spin samples *before* cutting the plastic strips to avoid spilling the PCR reaction.
18. Although other gel loading dyes can be used, we avoid using a dye containing bromophenol blue. Given than it migrates at approximately 150 bp on a 2% TBE agarose gel, which coincides with the library PCR product length, it makes proper gel excision of the library difficult. Orange G migrates at approximately 30 bp on a 2% TBE agarose gel.
19. "Empty" amplification products run at approximately 130 nt in length, thus it is important to properly resolve the gel for library gel-excision.
20. As a rule of thumb, libraries need to be diluted 1:10 for fragment analysis.

Acknowledgments

This work was supported by the European Research Council (ERC-StG-338252) and the Austrian Science Fund (Y-733-B22 START, W-1207-B09, and SFB F43-22) to S. L. A. and a Boehringer Ingelheim Fonds PhD Fellowship to M. M. R.-P.

References

1. Machnicka MA, Milanowska K, Osman Oglou O et al (2013) MODOMICS: a database of RNA modification pathways—2013 update. Nucleic Acids Res 41:D262–D267
2. Norbury CJ (2013) Cytoplasmic RNA: a case of the tail wagging the dog. Nat Rev Mol Cell Biol 14:643–653
3. Li X, Xiong X, Yi C (2016) Epitranscriptome sequencing technologies: decoding RNA modifications. Nat Methods 14:23–31
4. Lee M, Kim B, Kim VN (2014) Emerging roles of RNA modification: m(6)A and U-Tail. Cell 158:980–987
5. Reimão-Pinto MM, Ignatova V, Burkard TR et al (2015) Uridylation of RNA hairpins by tailor confines the emergence of MicroRNAs in Drosophila. Mol Cell 59:203–216
6. Martin G, Keller W (2007) RNA-specific ribonucleotidyl transferases. RNA 13:1834–1849
7. Dreyfus M, Régnier P (2002) The poly(A) tail of mRNAs: bodyguard in eukaryotes, scavenger in bacteria. Cell 111:611–613
8. Houseley J, Tollervey D (2009) The many pathways of RNA degradation. Cell 136: 763–776
9. De Almeida C, Scheer H, Zuber H et al (2017) RNA uridylation: a key posttranscriptional modification shaping the coding and noncoding transcriptome, Wiley interdisciplinary reviews. RNA 1829:e1440–e1425
10. Ameres SL, Zamore PD (2013) Diversifying microRNA sequence and function. Nat Rev Mol Cell Biol 14:475–488
11. Li J, Yang Z, Yu B, Liu J, Chen X. 2005. Methylation protects miRNAs and siRNAs from a 3'-end uridylation activity in Arabidopsis. Curr Biol 15(16):1501–7
12. Kim Y-K, Heo I, Kim VN (2010) Modifications of small RNAs and their associated proteins. Cell 143:703–709
13. Bortolamiol-Becet D, Hu F, Jee D et al (2015) Selective suppression of the splicing-mediated MicroRNA pathway by the terminal Uridyltransferase Tailor. Mol Cell 59:1–13
14. Lau NC, Lim LP, Weinstein EG et al (2001) An abundant class of tiny RNAs with probable regulatory roles in Caenorhabditis elegans. Science 294:858–862
15. Hafner M, Landgraf P, Ludwig J et al (2008) Identification of microRNAs and other small regulatory RNAs using cDNA library sequencing, methods (San Diego). CAL 44:3–12
16. Pfeffer S, Lagos-Quintana M, Tuschl T (2005) Cloning of small RNA molecules, current protocols in molecular biology/edited by Frederick M. Ausubel ... [et al.]. Chapter 26. Unitas 26:4–26.4.18
17. Vigneault F, Ter-Ovanesyan D, Alon S et al (2012) High-throughput multiplex sequencing of miRNA. Curr Protoc Hum Genet 11:Unit 11.12.1–Unit 11.1210
18. Heo I, Joo C, Cho J et al (2008) Lin28 mediates the terminal uridylation of let-7 precursor MicroRNA. Mol Cell 32:276–284
19. Hagan JP, Piskounova E, Gregory RI (2009) Lin28 recruits the TUTase Zcchc11 to inhibit let-7 maturation in mouse embryonic stem cells. Nat Struct Mol Biol 16:1021–1025
20. Heo I, Joo C, Kim Y-K et al (2009) TUT4 in concert with Lin28 suppresses microRNA biogenesis through pre-microRNA uridylation. Cell 138:696–708
21. Heo I, Ha M, Lim J et al (2012) Mono-uridylation of pre-MicroRNA as a key step in the biogenesis of group II let-7 MicroRNAs. Cell 151:521–532
22. Lehrbach NJ, Armisen J, Lightfoot HL et al (2009) LIN-28 and the poly(U) polymerase PUP-2 regulate let-7 microRNA processing in Caenorhabditis elegans. Nat Struct Mol Biol 16:1016–1020
23. Liu X, Zheng Q, Vrettos N et al (2014) A MicroRNA precursor surveillance system in quality control of MicroRNA synthesis. Mol Cell 55:868–879
24. Newman MA, Mani V, Hammond SM (2011) Deep sequencing of microRNA precursors reveals extensive 3′ end modification. RNA 17:1795–1803
25. Kim B, Ha M, Loeff L et al (2015) TUT7 controls the fate of precursor microRNAs by using three different uridylation mechanisms. EMBO J 34:1801–1815
26. Alon S, Vigneault F, Eminaga S et al (2011) Barcoding bias in high-throughput multiplex sequencing of miRNA. Genome Res 21:1506–1511
27. Chou M-T, Han BW, Hsiao C-P et al (2015) Tailor: a computational framework for detecting non-templated tailing of small silencing RNAs. Nucleic Acids Res 43:e109
28. Lopez-Gomollon S, Nicolas FE (2013) Purification of DNA Oligos by denaturing polyacrylamide gel electrophoresis (PAGE). Methods Enzymol 529:65–83
29. Baran-Gale J, Kurtz CL, Erdos MR et al (2015) Addressing Bias in small RNA library preparation for sequencing: a new protocol

recovers MicroRNAs that evade capture by current methods. Front Genet 6:352
30. Tian G, Yin X, Luo H et al (2010) Sequencing bias: comparison of different protocols of microRNA library construction. BMC Biotechnol 10:64
31. Stage DE, Eickbush TH (2007) Sequence variation within the rRNA gene loci of 12 Drosophila species. Genome Res 17:1888–1897
32. Seitz H, Ghildiyal M, Zamore PD (2008) Argonaute loading improves the 5′ precision of both MicroRNAs and their miRNA* strands in flies. Curr Biol 18:147–151
33. Jayaprakash AD, Jabado O, Brown BD et al (2011) Identification and remediation of biases in the activity of RNA ligases in small-RNA deep sequencing. Nucleic Acids Res 39:e141
34. Zhuang F, Fuchs RT, Sun Z et al (2012) Structural bias in T4 RNA ligase-mediated 3′-adapter ligation. Nucleic Acids Res 40:e54
35. Fuchs RT, Sun Z, Zhuang F et al (2015) Bias in ligation-based small RNA sequencing library construction is determined by adaptor and RNA structure. PLoS One 10:e0126049
36. Hafner M, Renwick N, Brown M et al (2011) RNA-ligase-dependent biases in miRNA representation in deep-sequenced small RNA cDNA libraries. RNA 17:1697–1712
37. Van Nieuwerburgh F, Soetaert S, Podshivalova K et al (2011) Quantitative bias in Illumina TruSeq and a novel post amplification barcoding strategy for multiplexed DNA and small RNA deep sequencing. PLoS One 6:e26969
38. Kozomara A, Griffiths-Jones S (2014) miRBase: annotating high confidence microRNAs using deep sequencing data. Nucleic Acids Res 42:D68–D73
39. Reimão-Pinto MM, Manzenreither RA, Burkard TR et al (2016) Molecular basis for cytoplasmic RNA surveillance by uridylation-triggered decay in Drosophila. EMBO J 35:2417–2434
40. Beckert B, Masquida B (2011) Synthesis of RNA by in vitro transcription. Methods Mol Biol 703:29–41
41. Kim Y-K, Yeo J, Ha M et al (2011) Cell adhesion-dependent control of microRNA decay. Mol Cell 43:1005–1014

Chapter 11

Uncovering the Stability of Mature miRNAs by 4-Thio-Uridine Metabolic Labeling

Matteo J. Marzi and Francesco Nicassio

Abstract

MicroRNAs (miRNAs) are an evolutionary conserved class of short, single-stranded noncoding RNAs (<18–22 nt in length) that act in posttranscriptional regulation of gene expression in higher eukaryotes. The abundance of a miRNA is a key feature in control of its activity and, therefore, a number of mechanisms finely regulate miRNA levels, acting at both transcriptional and posttranscriptional level. Recent evidences, including our research, highlighted the role of miRNA decay as a mechanism controlling the miRNA pool. We describe in this chapter an optimized methodology to determine miRNA degradation rates in mammalian cells. Our approach is based on metabolic pulse labeling with 4-thiouridine (4sU), a uridine analog that is incorporated in nascent RNA and allows thiol-specific biotinylation and selective pull-down of labeled RNA. In particular, given the long average half-life and the complex biogenetic process of miRNAs, we developed a "pulse-chase" protocol where 4sU is removed from the medium after a long labeling period (2–3 h pulse), and labeled RNA is purified at different time points to measure the decay of labeled molecules. By combining the 4sU-based "pulse-chase" approach with high-throughput small RNA sequencing (sRNAseq), it is possible to quantify at genome-wide level miRNA degradation rates.

Key words microRNAs, Small RNAs, Half-life, Decay, Degradation, Turnover, sRNAseq, Sequencing, 4-thio-uridine, 4sU, Pulse, Chase

1 Introduction

While in the past miRNAs were thought to be a homogenous class of stable molecules with very slow turnover rates (more than 24 h, or even weeks), recent studies revealed that miRNAs turnover in mammalian cells is rather heterogeneous, with many species with a relatively fast turnover (<14 h) as opposed to other miRNAs with a stable behavior [1–3]. Previous attempts to measure globally the degradation rates of miRNAs relied on the use of inhibitors of miRNA transcription or processing, but these approaches dramatically alter the physiology of the cells [4]. An alternative method to monitor RNA synthesis and decay is via metabolic labeling: endogenous RNAs are labeled by nucleotide analogs [e.g., 4-thiouridine (4sU), 5-etyniluridine (EU) and 5′-bromo-uridine (BrU), analogs

Ulf Andersson Vang Ørom (ed.), *miRNA Biogenesis: Methods and Protocols*, Methods in Molecular Biology, vol. 1823,
https://doi.org/10.1007/978-1-4939-8624-8_11,

of uridine], followed by purification of de novo labeled RNAs [5]. Among the others, 4sU is rapidly taken up by cells and phosphorylated by uridine kinases. In mouse fibroblasts, administration of 4sU at 300 μM concentration and for up to 3 h does not affect cell proliferation or alter global miRNA expression [6, 7]. After labeling, it is possible to separate nascent and pre-existing untagged RNA using thiol-specific biotinylation and affinity purification with streptavidin-coated magnetic separation [8]. For mRNA, the half-life can be calculated from the newly transcribed RNA/total RNA ratios (R) and the duration of labeling (t_L) as $t_{1/2} = t_L \times \ln(2)/\ln(1-R)$. Nonetheless, miRNA half-life is better derived by chasing decreasing levels of labeled RNA [2]. This is due to the complex biosynthesis of miRNAs, that are transcribed as precursors (pri-miRNA, which incorporates 4sU in first instance) and originate from two sequential processing events [4]. In addition, the use of a relatively short labeling time is not appropriate to model half-life of RNA species with high stability, such as most miRNAs. To circumvent these issues, we developed a pulse-chase approach (Fig. 1) in which we label newly synthesized RNA with 4sU for 2–3 h (a time sufficient to complete the entire biogenetic process of miRNAs) and then we follow the decay of labeled RNA for up to 16 h. To measure miRNAs, we employed RT-qPCR (if few miRNAs have to be analyzed) or small RNA sequencing by Illumina technology (if all miRnome is under investigation). In the latter case, reads are analyzed with the IsomiRage tool [9] to provide a quantification of both mature miRNAs and their variants (IsomiRs), which can correlate with the decay process [4]. Finally, we provide a normalization approach and indications on the best way to model miRNA degradation through a one-phase or two-phase exponential decay.

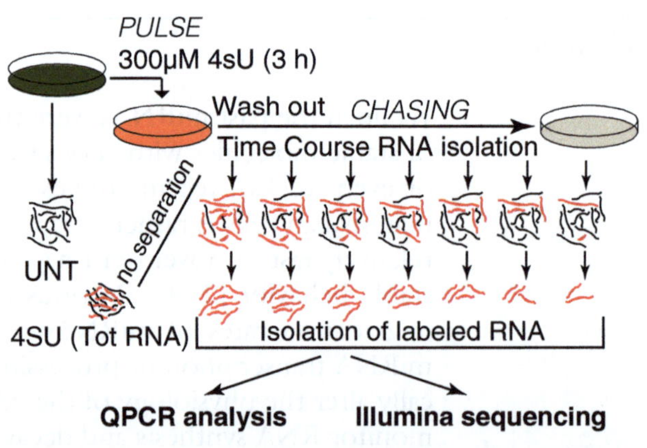

Fig. 1 4sU pulse-chase experimental strategy. The scheme illustrates the experimental strategy used to measure miRNA decay by a 4sU pulse-chase protocol. After labeling (pulse of 4sU), samples are chased with unlabeled uridine and collected at different time points (up to 24 h). Labeled RNA can be isolated from total RNA and used in downstream applications

2 Materials

Prepare all solutions in nuclease-free water, such as commercial DEPC-treated ultra-pure water.

2.1 Pulse and Chase Experiment by Metabolic Labeling

1. 4-thiouridine: 75 mM stock is prepared by dissolving 100 mg in 5123 mL of sterile water in a laminar flow hood. Store 4sU as 200 μL aliquots at −20 °C. Discard unused thawed reagent.
2. Uridine: 75 mM stock is prepared by dissolving 100 mg in 5460 mL of sterile water in a laminar flow hood. Store as 200 μL aliquots at −20 °C.
3. Cell Medium: Preferred growth medium for cells of choice.
4. Sterile PBS.

2.2 Isolation of Labeled RNA

1. Trizol, or equivalent
2. Chloroform
3. Ethanol
4. Isopropanol
5. Spin-column based purification kit (*see* **Note 1**)
6. Nuclease-free water

2.3 Biotinylation and Removal of Unbound Biotin-HPDP

1. Biotin Labeling Buffer (2.5× stock): 25 mM Tris-HCl (pH 7.4) and 2.5 mM EDTA in water. Prepare using nuclease-free solutions.
2. Biotin-HPDP/DMF stock (1 mg/mL): 50 mL dimethylformamide (DMF), 50 mg biotin-HPDP (*see* **Note 2**). Work at fume hood, mixing thoroughly and heating at 37 °C for 1 h or until dissolved. Store 1 mL aliquots at −80 °C.
3. MaXtract high density tubes or equivalent.
4. 5 M NaCl.

2.4 Isolation of 4sU-Labeled RNA Using Streptavidin-Beads

1. Dynabeads MyOne Streptavidin T1
2. Dynabeads Washing solution A: 100 mM NaOH, 50 mM NacL
3. Dynabeads Washing solution B: 100 mM NaCl
4. Dynabeads Washing Buffer 2×: 2 M NaCl, 10 mM Tris-HCl (pH 7.5), 1 mM EDTA, 0.1% Tween-20
5. Elution solution: 10 mM EDTA in 95% formamide (*see* **Note 3**)

2.5 Final Purification of 4sU-Labeled RNA

1. Trizol-LS reagent (*see* **Note 4**)
2. miRNA purification kit (*see* **Note 1**)
3. Qubit RNA HS Assay Kit or Nanodrop Spectrophotometer

3 Methods

Carry out all procedures at room temperature unless otherwise specified.

3.1 Pulse and Chase Experiment by Metabolic Labeling

1. Plate cells (*see* **Note 5**) the day before the experiment in 15 cm dishes. Aim to achieve 60–70% confluent cells at the start of labeling (see **Note 6**). Typically, you will require at least 7 dishes to perform a single pulse-chase experiment (*see* **Note 7** and Fig. 1).
2. Thaw 4sU (75 mM) just before use and dilute 1:250 in cell culture medium to a final concentration of 300 μM, or the appropriate concentration established in your cellular model (*see* **Note 8**). Add the 4sU-containing medium to the cells.
3. Incubate for the required time (typically 2–3 h, *see* **Note 5**).
4. Thaw Uridine (75 mM) and dilute 1:75 in cell culture medium to a final concentration of 1 mM.
5. Remove 4sU-containing medium, wash 2× with PBS, and replace with Uridine-containing medium.
6. At desired time points, remove cell culture medium from cells, wash twice with cold PBS on ice, and add 1 mL Trizol to each plate. Collect lysed cells with a scraper. Samples can be stored at −80 °C for at least 1 month.

3.2 Isolation of Labeled RNA

1. Incubate samples (i.e., in 1 mL of Trizol) for 5′ at RT.
2. Add 200 μL of chloroform and shake the tube vigorously for 15″.
3. Incubate for 2–3′ at RT.
4. Centrifuge for 15′ at 12,000 × *g* at 4 °C.
5. Transfer the upper aqueous phase to a new tube.
6. Add 1.5 volumes of 100% EtOH (*see* **Note 9**) and mix by pipetting.
7. Pipet (700 μL at a time) into a spin column (*see* **Note 10**: capacity 100 μg), centrifuge ≥8,000 × *g* for 15″ at RT.
8. Pipet remaining sample into the column, centrifuge ≥8,000 × *g*, 15″ at RT.
9. Perform the on-column DNase digestion with the RNase-Free Dnase Set.

   (a) Add 350 μL of RWT to the column, centrifuge 15″ at ≥8,000 × g

   (b) Add 10 μL DNaseI to 70 μL of buffer RDD.

(c) Apply 80 μL to the center of the column and incubate 15′ at RT.

(d) Add 350 μL of RWT to the column, centrifuge 15″ at ≥8,000 × g.

10. Wash with 500 μL Buffer RPE and centrifuge 2′ at ≥8,000 × g.
11. Place the column in a new collection tube, centrifuge at full speed for 1′
12. Place the column in a 1.5 collection tube and pipet 50 μL of RNase-free water in the center of the column. Centrifuge 1′ at ≥8,000 × g.
13. Repeat with a second volume of 50 μL RNase-free water.

3.3 Biotinylation of 4sU-Labeled RNA

1. Typically start with 30–50 μg of total RNA, adjust to 100 μL with RNase-free water. It is advisable to introduce appropriate positive and negative controls (*see* **Note 11**).
2. Add 100 μL of biotin labeling buffer (2.5×).
3. Add 50 μL of Biotin-HPDP/DMF at fume hood and mix by pipetting.
4. Incubate for 2 h with rotation.
5. Centrifuge MaXtract columns: 12,000–16,000 × *g* for 20–30 s.
6. Add chloroform/isoamyl alcohol (24:1) at equal volume (250 μL) to biotinylation samples and mix.
7. Add samples into the MaXtract tubes and mix phases thoroughly by repeated inversion (do not vortex).
8. Centrifuge at 12,000–16,000 × *g* for 5 min.
9. Take the upper, RNA-containing, aqueous phase into a fresh tube (about 220 μL).
10. Add 5 M NaCl at 1/10 volume (22 μL) (*see* **Note 12**).
11. Add isopropanol at an equal volume (220 μL), mix thoroughly by pipetting.
12. Precipitate at full speed for 30 min at 4 °C.
13. Discard the liquid.
14. Add 75% ethanol at an equal volume (220 μL).
15. Precipitate at full speed for 10 min.
16. Dry the pellet at RT (do not overdry).
17. Resuspend pellet in 100 μL RNase-free water. To completely dissolve and denature RNA, place the tubes at 65 °C for 10′ in a shaker and then immediately place on ice (*see* **Note 13**).

3.4 Isolation of 4sU-Labeled RNA Using Streptavidin-Beads

1. Wash Dynabeads MyOne Streptavidin T1:

(a) Resuspend beads by vortexing.

(b) Take 50 μL beads/sample into a new tube.

(c) Place the tube on a magnet for 2′.
(d) Remove supernatant with a pipette while the tube is on the magnet.
(e) Remove the tube from the magnet.
(f) Wash beads twice with solution A (2′), using the same initial beads volume.
(g) Wash beads once in solution B (2′), using the same initial volume.
(h) Resuspend beads in 2× washing buffer, using twice the original volume (final concentration is 5 μg/μL).

2. Add an equal volume of washed Dynabeads (100 μL) to biotinylated RNA in H_2O (dilutes the 2× Dynabeads washing buffer to 1×).
3. Incubate for 15′, room temperature, gentle rotation (e.g., on a carrousel).
4. Separate beads with magnet for 2′ (Flow-through can be collected for unlabeled pre-existing RNA recovery; *see* **Note 14**).
5. Dilute Dynabeads washing buffer 2× to 1× for the required volume.
6. Wash 3× with 200 μL of Dynabeads washing buffer (washing fractions can be collected for RNA recovery; *see* **Note 14**). Keep the beads on the magnet during washings.
7. Add 100 μL of 10 mM EDTA in 95% formamide to elute labeled RNA: resuspend the beads by pipetting and incubate 5′–10′ at 65 °C.
8. Separate beads with magnet for 2′, take supernatant into a new tube.
9. Clean again the supernatant on the magnet to remove residual beads.

3.5 Final Purification of 4sU-Labeled RNA

1. Start with the samples eluted from the beads (100 μL in EDTA/formamide.
2. Add 100 μL of RNase-free water and 600 μL of Trizol-LS (*see* **Note 4**), mix well, and leave 5′ at RT.
3. Add 160 μL of chloroform and mix well by hand shaking for 15″.
4. Incubate for 2–3′ at RT.
5. Centrifuge for 15′ at 12,000 × *g* at 4 °C.
6. Transfer the upper aqueous phase to a new tube.
7. Add 1.5 volumes of 100% EtOH and mix by pipetting.
8. Pipet (700 μL at a time) into a spin micro-column (*see* **Note 1**).
9. Centrifuge ≥8,000 × *g* for 15″ at RT.

10. Pipet remaining sample (if any) into the column, centrifuge ≥8,000 × *g* for 15″ at RT.
11. Wash with 700 μL of RWT to the column, centrifuge 15″ at ≥8,000 × *g*.
12. Wash with 500 μL Buffer RPE and centrifuge 2′ at ≥8,000 × *g*.
13. Wash with 500 μL of 80% Ethanol and centrifuge 2′ at ≥8,000 × *g*.
14. Place the column in a new collection tube, centrifuge at full speed for 5′.
15. Place the column in a 1.5 collection tube and pipet 14 μL of RNase-free water in the center of the column. Centrifuge 1′ at ≥8,000 × *g*.
16. Quantify 1 μL of eluted RNA (*see* **Note 15**).

3.6 RT-QPCR Validation and HT-Sequencing of Isolated 4sU-Labeled miRNA

Downstream applications of 4sU-labeled RNA include the analysis by RT-qPCR of specific genes or the genome-wide high-throughput sequencing of the entire transcriptome/miRNome. Routinely, we first measure a short-lived transcript (usually Myc or primary miRNA precursors) and a long-lived one (usually Gapdh) in both total and 4su-RNA to validate the experiment. We recommend using 2 μL (i.e., 1/7 of the total, see last step of Subheading 3.5) of 4sU-RNA for cDNA synthesis (20 μL mix). Then, dilute 1:10 the cDNA and use 5 μL of diluted cDNA for qPCR reactions. RNA half-lives can be determined from the qPCR results by linear regression analysis (Fig. 2a). To monitor miRNA decay by RT-qPCR, we employ a miRNA-specific RT-qPCR kits. Given the context-dependency of miRNA degradation rates [2, 3, 10], it might be difficult to find known short-lived and long-lived miRNAs available for your cellular model. Usually, miRNAs like miR-182-5p and miR-503-5p have a quick turnover (e.g., fibroblasts and neurons) [1–3]. Conversely, miRNAs like miR-21, miR-25, or miR-92a have an extremely long half-life in fibroblasts [2]. The normalization of the results might be problematic. It is possible to use as housekeeping genes other small RNA species, such as snoRNAs, but also their decay is not well characterized. We prefer using a stable miRNA (e.g., miR-92a-3p) for normalization, to correct also for biases in miRNA recovery or miRNA biogenesis.

After having checked the quality of the experiment by RT-qPCR, small RNA sequencing can be performed, starting from at least 200 ng of 4sU-labeled RNA. It is also advisable to check for any aspecific effect of 4sU labeling on global miRNA expression, especially if working on cell lines on which 4sU labeling of miRNA has not been tested yet. For instance, we compare small RNAs profiles obtained from untreated cells (not labeled without isolation—UNT) and from total RNA pulsed with 4sU (labeled without isolation—4sU). The expression of miRNAs should be

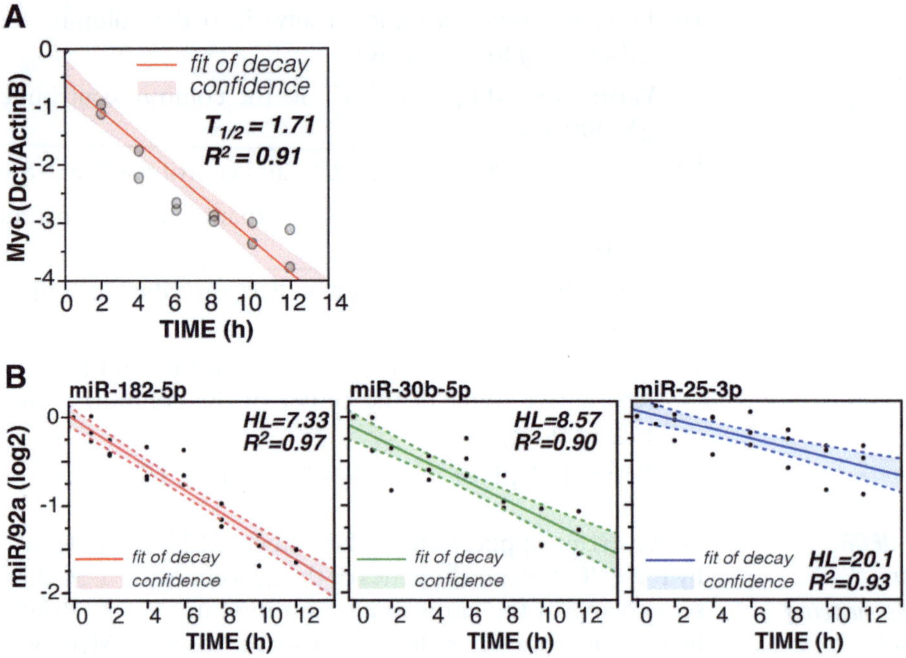

Fig. 2 Modeling RNA half-lives from pulse-chase experiments. (**a**) The half-life of Myc was measured by RT-qPCR using the pulse-chase protocol performed in mouse 3T9 fibroblasts. Ct values at different time points were first normalized to a stable mRNA (ActinB) and then DDCt were calculated over the 0 h time point (i.e., the end of the pulse). Two independent experiments (shown as dots) were performed and the average DDCt was considered for the fitting. Decay was modeled according to a one-phase exponential fitting. (**b**) Examples of miRNA half-lives measurements by small RNA sequencing. Raw reads for each miRNAs were log2 transformed, normalized (using miR-92a-3p as housekeeping), and reported as log2 fold change over the 0 h time point. Three independent experiments were performed and average values were used for the fitting, performed by one-phase exponential decay in GraphPad Prism6. The plateau was fixed to a constant value of zero and the fitting was based on least squares. Although in the shown examples the fitting was good also for a miRNA with a long half-life (miR-25-3p), it should be noted that this experiment is best suited to monitor decay rates comprised between 2 and 12 h

similar in the two conditions with a correlation of reads close to perfect linear correlation (R^2 = 0.98–1.00). A miRNA which is reproducibly differentially expressed upon 4sU treatment should be discarded in the following analyses. We perform sequencing of small RNAs on an Illumina HiSeq 2000 platform at 50 bp single-read mode and with 20 million read depth (8×). Data analysis is performed with the *IsomiRage* workflow, as previously described [9]. We strongly suggest employing a bioinformatics pipeline that keeps into account miRNA isoforms, since, when degraded, miRNAs usually accumulate modifications [tailing (addition of non-templated nucleotides at the 3′ end) and trimming (shortening)]. We previously showed that 3′ non-templated isomiRs are enriched in fast-decaying miRNAs [2].

It is important to remind that, during the time-course, cells undergo mitotic divisions, leading to dilution of 4sU-labeled miR-

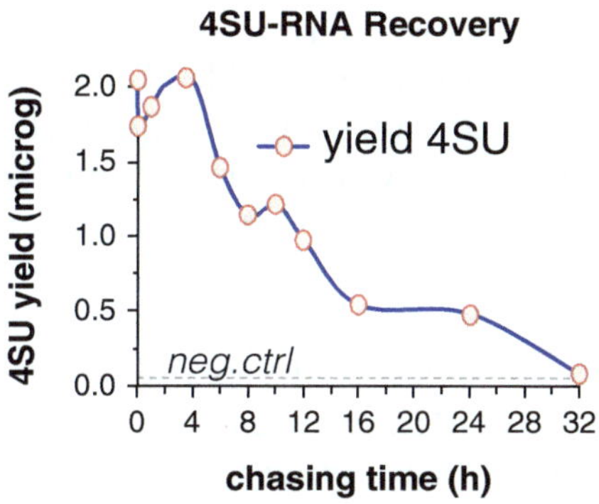

Fig. 3 4sU-RNA yield during chasing. The yield of 4sU-RNA recovery (from 40 μg of total RNA) was calculated in 3T9 mouse fibroblasts at different time points of chasing. Please note that in the first 2 h of chasing there is no reduction in the 4sU-yield, suggesting that, in this time window, 4sU is still available in the cell. Note: 3T9 cells have a doubling time of 18 h, thus every hour 4sU is diluted by a log2 factor of −0.056 (−1/18), in addition to the decrease due to degradation

NAs. Thus, also stable species apparently decrease over time. Thus, normalization of RNA content by cell division is a critical step for data interpretation. In our previous work using 3T9 fibroblasts, data were normalized using a single stable miRNA (miR-92a-3p, considering both canonical miRNA reads and templated reads but excluding other isomiRs), or the geomean of multiple stable miRNA. Library size normalization is not suggested (having the effect to introduce artifacts, such as miRNA increasing over time during the chase), but it might be used in absence of other options to isolate qualitatively the group of miRNAs with a faster or slower decay (without an exact quantification of the real half-life). For accurate modeling of half-lives, it is highly recommended to perform at least three independent time-course experiments, with 6–8 time points. Half-life can be modeled, from normalized data, using a one-phase exponential decay function (Fig. 2b), such as the one provided in Prism (GraphPad). Although it is not possible to anticipate the results of other cell models, it is expected that passenger miRNAs will display an overall shorter half-life than guide miRNAs [2, 11]. Finally, it is relevant to stress that the pulse-chase protocol is particularly suited to measure decay rates up to 12–16 h, a time when 4sU-labeled RNA is still sufficient for small RNA sequencing (Fig. 3). Thus, for most highly stable miRNAs (>20 h) the measurement of half-lives tends to be more inaccurate.

4 Notes

1. *Isolation of total RNA including miRNAs*: In this chapter, we included the commercial kits that we routinely use in our lab to perform the isolation of 4sU-labeled miRNAs. Obviously, it is possible to adopt other solutions, but it is important to con-

trol before starting that miRNA recovery is efficient and extremely reproducible. The maximum yield of a spin column is usually 100 μg: if starting with higher amounts, it is important to use more than one column. The final yield of 4sU-RNA is usually <1 μg. It was reported that isolating miRNAs from <1 μg RNA by precipitation is highly inefficient [12], thus for the final step of the procedure (Subheading 3.5) please adopt a column-based isolation method.

2. *Use of MTS-biotin in replacement of HPDP-biotin*: It was published that biotinylation with HPDP-biotin can be inefficient and introduce a length-bias in sequencing experiments [11]. We performed in-house experiments to directly compare the two biotins and we found that MTS-biotin has indeed a better yield (~3 times higher than HPDP-biotin) but also a much higher background (~10 times higher than HPDP-biotin), when following the proposed protocol. Thus, we discourage the use of MTS-biotin with our pulse-chase procedure, unless the yield is a limiting factor in your cellular model.
3. *DTT vs. formamide.* Once the interaction of streptavidin and biotin is formed, it is unaffected by wide extremes of pH, temperature, and denaturing agents. 4sU-labeled RNA can be eluted with 98% efficiency by adding 10 mM EDTA/95% formamide and incubating at 65 °C for at least 2′. Alternatively, elute by adding 100 μL of 100 mM dithiothreitol (DTT) [6]. Repeat the elution twice. Prepare DTT in nuclease-free water.
4. *Trizol-LS*: This formulation of Trizol is specifically designed to isolate RNA from liquid samples. Alternatively, it is possible, though not suggested, to add 0.9 mL of normal Trizol to 100 μL of EDTA/formamide eluted samples.
5. *Cell lines.* We routinely performed 4sU pulse-chase experiments in 3T9 mouse fibroblasts. In this cellular model, we tested extensively the dynamics of 4sU during the pulse-chase (see Fig. 3). We also tested other cell lines, including human cancer cells (HeLa) and human foreskin fibroblasts (BJ). In general, the protocol requires a large amount of RNA (>40 μg). Thus, it might be difficult to perform the pulse-chase experiments on cell types difficult to be grown in large quantities (e.g., primary cells or rare sub-populations). Before starting, it is important to adjust 4sU concentration and labeling duration to the employed cell line. First, we suggest labeling cells with increasing 4sU-concentrations at a fixed incorporation time (using 2 h as a starting point). Select the concentration at which purified newly transcribed RNA reaches a plateau. Then, using the defined dose, perform a chasing experiment and monitor the 4sU yield (note: it is possible that, during the first 2 h of chasing, 4sU yield will not decrease; this is probably due to residual 4sU in the system). At this stage, check carefully

that no toxicity is observed along the entire time-course, especially if high dose of 4sU (>300 μM) has been employed.

6. *Cell density.* Cell density is a critical factor, regulating both 4sU incorporation and microRNAs processing [13]. Thus, to avoid this confounding factor when measuring miRNA half-lives, the pulse-chase experiment should be performed either in confluent cells or in not confluent cells, but never mixing the two conditions. Thus, cell culture plates should be checked carefully (density should be <70%) before starting the labeling.
7. *Experimental layout.* We suggest to plate two 15 cm plates for each time point, collecting samples every two hours from the end of labeling (time 0) up to 12 h of chasing (i.e., 14 plates should be included for each experiment). For a scheme of the samples required for a typical experiment, refer to Fig. 1.
8. *Light Exposure.* 4sU is a photoactivatable ribonucleoside and 4sU-containing RNA is efficiently crosslinked to proteins after exposure to 365 nm light source. 4sU-treated cells should be cultured in the dark and exposure to bright light should be avoided.
9. *Ethanol.* MiRNA isolation with spin-column methods requires the use of 1.5 volumes of 100% ethanol. Typically, the isolation of longer RNA species is performed using 1 volume of 70% ethanol (e.g., RNeasy mini kit). If using a kit not specifically designed to isolate miRNAs, please change the ethanol concentration at this step.
10. *Spin-column Capacity.* If starting with >100 μg total RNA expected yield, load the sample on two spin-columns.
11. *Positive and negative controls.* It is advisable to include (in each 4sU purification) a positive and negative control. The positive control is a total RNA from labeled cells that was previously successfully purified, while the negative is a total RNA from non-labeled cells. We usually prepare large amounts (>200 μg) of positive and negative samples before starting the pulse-chase experiments. 4sU yields and purity are then monitored during the entire duration of the study with the same control set.
12. *Recovery of biotinylated RNA.* It is important to note that spin-column based kits should not be used to recover biotinylated RNA as they contain reducing agents, which cleave the disulfide bond and remove the biotin from 4sU-RNA.
13. *Biotinylated-RNA yield.* Before proceeding with streptavidin isolation, quantify precipitated biotinylated-RNA (take care that RNA is completely dissolved before measuring the concentration). We usually recover 90% of the starting material (e.g., 36 μg from 40 μg). Monitor in particular that no major differences in recovery are present across samples. Yields at this stage might be used to normalize the final 4sU/total RNA ratio.

14. *Unlabeled RNA*. Recovery on unlabeled RNA is optional and can be performed by collecting the flow-through (unbound RNA after adding the RNA-streptavidin beads solution to the magnet) and the first wash with Washing Solution 1×. Purification should be done by NaCl and Isopropanol precipitation as detailed in Subheading 3.3, **steps 10–17**.
15. *Dosing 4sU-RNA and expected results*. Eluted 4sU-RNA can be dosed using a Nanodrop Spectrophotometer or a Qubit fluorometric quantitation. When labeling for 2–3 h, it is typical to get 1–5% newly transcribed RNA/total RNA (i.e., 400–2000 ng of 4sU-RNA yield). Nonetheless, the yield decreases with chasing time (Fig. 3). If the measured concentration is <20 ng/μL, we strongly recommend to use the Qubit assay.

References

1. Krol J, Busskamp V, Markiewicz I, Stadler MB, Ribi S, Richter J, Duebel J, Bicker S, Fehling HJ, Schübeler D (2010) Characterizing light-regulated retinal microRNAs reveals rapid turnover as a common property of neuronal microRNAs. Cell 141(4):618–631. https://doi.org/10.1016/j.cell.2010.03.039
2. Marzi MJ, Ghini F, Cerruti B, de Pretis S, Bonetti P, Giacomelli C, Gorski MG, Kress T, Pelizzola M, Muller H, Amati B, Nicassio F (2016) Degradation dynamics of microRNAs revealed by a novel pulse-chase approach. Genome Res 26:554. https://doi.org/10.1101/gr.198788.115
3. Rissland OS, Hong S-J, Bartel DP (2011) MicroRNA destabilization enables dynamic regulation of the miR-16 family in response to cell-cycle changes. Mol Cell 43(6):993–1004. https://doi.org/10.1016/j.molcel.2011.08.021
4. Krol J, Loedige I, Filipowicz W (2010) The widespread regulation of microRNA biogenesis, function and decay. Nat Rev Genet 11(9):597–610. https://doi.org/10.1038/nrg2843
5. Tani H, Akimitsu N (2012) Genome-wide technology for determining RNA stability in mammalian cells: historical perspective and recent advantages based on modified nucleotide labeling. RNA Biol 9(10):1233–1238. https://doi.org/10.4161/rna.22036
6. Rabani M, Levin JZ, Fan L, Adiconis X, Raychowdhury R, Garber M, Gnirke A, Nusbaum C, Hacohen N, Friedman N, Amit I, Regev A (2011) Metabolic labeling of RNA uncovers principles of RNA production and degradation dynamics in mammalian cells. Nat Biotechnol 29(5):436–442. https://doi.org/10.1038/nbt.1861
7. Marzi MJ, Puggioni EM, Dall'Olio V, Bucci G, Bernard L, Bianchi F, Crescenzi M, Di Fiore PP, Nicassio F (2012) Differentiation-associated microRNAs antagonize the Rb-E2F pathway to restrict proliferation. J Cell Biol 199(1):77–95. https://doi.org/10.1083/jcb.201206033
8. Dolken L, Ruzsics Z, Radle B, Friedel CC, Zimmer R, Mages J, Hoffmann R, Dickinson P, Forster T, Ghazal P, Koszinowski UH (2008) High-resolution gene expression profiling for simultaneous kinetic parameter analysis of RNA synthesis and decay. RNA 14(9):1959–1972. https://doi.org/10.1261/rna.1136108
9. Muller H, Marzi MJ, Nicassio F (2014) IsomiRage: from functional classification to differential expression of miRNA isoforms. Front Bioeng Biotechnol 2:38. https://doi.org/10.3389/fbioe.2014.00038
10. de la Mata M, Gaidatzis D, Vitanescu M, Stadler MB, Wentzel C, Scheiffele P, Filipowicz W, Grosshans H (2015) Potent degradation of neuronal miRNAs induced by highly complementary targets. EMBO Rep 16(4):500–511. https://doi.org/10.15252/embr.201540078
11. Duffy EE, Rutenberg-Schoenberg M, Stark CD, Kitchen RR, Gerstein MB, Simon MD (2015) Tracking distinct RNA populations using efficient and reversible covalent chemistry. Mol Cell 59(5):858–866. https://doi.org/10.1016/j.molcel.2015.07.023
12. Kim YK, Yeo J, Kim B, Ha M, Kim VN (2012) Short structured RNAs with low GC content are selectively lost during extraction from a small number of cells. Mol Cell 46(6):893–895. https://doi.org/10.1016/j.molcel.2012.05.036
13. Hwang HW, Wentzel EA, Mendell JT (2009) Cell-cell contact globally activates microRNA biogenesis. Proc Natl Acad Sci U S A 106(17):7016–7021. https://doi.org/10.1073/pnas.0811523106

Chapter 12

Detection of microRNA-Target Interactions by Chimera PCR (ChimP)

James P. Broughton and Amy E. Pasquinelli

Abstract

MicroRNAs (miRNAs) regulate gene expression by directing Argonaute proteins to target RNAs, which usually results in destabilization and translational inhibition of the target RNA. The prediction of animal miRNA target sites has remained a challenge due to the ability of miRNAs to bind target RNAs through imperfect base pairing. Recently, several labs have established methods to produce biochemical evidence of miRNA-target interactions by generating chimeric reads where the miRNA is ligated to its target RNA. Despite the insights that can be gained from chimera producing methods, the current approaches are inefficient, labor intensive and require computational expertise. Here we describe a method, called Chimera PCR (ChimP), for the validation or testing of specific miRNA-target interactions. This method allows for focused experiments to analyze miRNA targeting in a variety of conditions.

Key words microRNA, ALG-1, Chimeric reads, Argonaute, AGO

1 Introduction

MicroRNAs (miRNAs) have emerged as essential regulators of development, cell differentiation, and stress responses in many eukaryotic systems. MiRNAs are small (~22 nucleotide) RNA molecules that bind and direct Argonaute (AGO) proteins to target RNAs by base-pairing to complementary sequences. In animals, miRNA targeting generally occurs in the 3′ untranslated region (3′UTR) of messenger RNAs (mRNAs) and leads to translational inhibition or transcript destabilization [1].

Despite the critical regulatory roles of miRNAs, a complete understanding of the rules of miRNA targeting remains elusive. Complicating the identification of miRNA targets is the ability of miRNAs to functionally regulate target genes through imperfect base pairing [2]. Therefore, an individual miRNA can potentially regulate hundreds of genes. To experimentally determine miRNA target sites, several groups have recently developed techniques that provide biochemical evidence of miRNA-target interactions [3–6].

Ulf Andersson Vang Ørom (ed.), *miRNA Biogenesis: Methods and Protocols*, Methods in Molecular Biology, vol. 1823, https://doi.org/10.1007/978-1-4939-8624-8_12, © Springer Science+Business Media, LLC, part of Springer Nature 2018

These methods are adapted from UV crosslinking and immunoprecipitation followed by sequencing (CLIP-seq), which identifies the direct interaction sites of RNA binding proteins [7]. Specifically, these new methods include steps during the "on-bead" biochemistry to ligate miRNAs to their target RNAs. The resulting hybrid RNA is called a chimeric read. The analysis of chimeric reads has generated expanded maps of miRNA binding in several systems and has led to new insights into how miRNAs recognize their target sites [3–6].

Chimera generating methods are suitable for identifying miRNA target sites globally. However, performing these methods generally requires the use of radioactivity and the analysis of chimeric reads is computationally challenging. In many cases, research groups are interested in how the binding of a limited number of miRNAs changes in various conditions on one or more target RNAs. To address this need for focused experiments, we developed a method called Chimera PCR (ChimP), which reveals miRNA-target interactions using PCR. ChimP allows for gene-specific experiments without the need for radioactivity or the requirement of complex bioinformatics analyses. Here we describe the application of the ChimP method to identify miRNA target sites for Argonaute-Like-Gene-1 (ALG-1) in the nematode worm *Caenorhabditis elegans* (*see* **Note 1**).

2 Materials

2.1 *C. elegans* Cultures, UV Crosslinking, and Homogenization

1. Materials for culturing *C. elegans* (OP50-seeded NGM agar plates, etc.)
2. M9 Buffer: 22 mM KH_2PO_4, 22 mM Na_2HPO_4, 85 mM NaCl, and 1 mM $MgSO_4$
3. UV Crosslinker
4. Standard NGM agar plates for *C. elegans* with no bacterial food on them
5. Homogenization buffer: 100 mM NaCl, 25 mM HEPES, 250 μM EDTA pH 8.0, 0.1% NP-40, 1 tablet/20 mL cOmplete Mini EDTA-free Protease Inhibitor Cocktail (Roche), 2 mM DTT, 25 U/mL rRNasin
6. Sonic Dismembrator (Model 100, Fisher Scientific)

2.2 ChimP

1. Protein G Dynabeads
2. RNase I storage buffer: 50 mM Tris-HCl pH 8.0, 100 mM NaCl, 0.01% Triton X-100, 50% glycerol
3. α-ALG-1 antibody (PA1-031, Thermo Fisher Scientific)
4. BCA Protein Assay Kit

5. RNase I
6. Turbo DNase
7. Thermomixer
8. High-salt wash buffer: 50 mM Tris-HCl pH 7.4, 1 M NaCl, 1 mM EDTA, 1% NP-40, 0.1% SDS, 0.5% sodium deoxycholate
9. Low-salt wash buffer: 20 mM Tris-HCl pH 7.4, 10 mM $MgCl_2$; 0.2% Tween-20
10. 10× PNK Buffer
11. 10 mM ATP
12. T4 PNK 3′ phosphatase minus
13. SUPERase In RNase Inhibitor
14. PEG400
15. 4× RNA Ligase Buffer: 200 mM Tris-HCl pH 7.4, 40 mM $MgCl_2$, 4 mM DTT
16. T4 RNA Ligase 1
17. 5× PNK pH 6.5 buffer: 350 mM Tris-HCl pH 6.5, 50 mM $MgCl_{2,}$ 5 mM DTT
18. Proteinase K
19. PK buffer: 100 mM Tris-HCl pH 7.4, 50 mM NaCl, 10 mM EDTA
20. Urea
21. Phenol:Chloroform:Isoamyl Alcohol (25:24:1)
22. Chloroform:Isoamyl Alcohol (24:1)
23. Glycogen
24. 100% Ethanol
25. 100% Isopropanol
26. 10 mM dNTPs
27. 5× First-strand RT Buffer
28. 0.1 M DTT
29. Superscript III
30. Agencourt AMPure XP beads
31. 10× CircLigase Buffer II
32. 50 mM MnCl2
33. CircLigase II
34. NEBuffer 3.1
35. BamHI
36. Accuprime Supermix I
37. 6% TBE gel

38. SYBR-GOLD
39. 6× Loading Dye
40. Phusion 2× HF Master Mix

2.3 Primer Sequences

1. L3: /5rApp/AGATCGGAAGAGCGGTTCAG/3ddc/ (IDT)
2. Rclip: /5Phos/NNTCCGNNNAGATCGGAAGAGCGTCGTGgatcCTGAACCGC (IDT)
3. cut_oligo_ddc: GTTCAGGATCCACGACGCTCTTC/3ddc/ (IDT)
4. P5 solexa: AATGATACGGCGACCACCGAGATCTACACTCTTTCCCTACACGACGCTCTTCCGATCT (IDT)
5. P3 solexa: CAAGCAGAAGACGGCATACGAGATCGGTCTCGGCATTCCTGCTGAACCGCTCTTCCGATCT (IDT)

3 Methods

3.1 Preparation of Lysates

1. Grow worms and collect at desired stage (need at least 50 μL packed worms, more is preferable as it makes later steps easier).
2. Wash worms (3×) with 10 mL M9 in 15 mL falcon tube. Resuspend in 5 mL M9.
3. Rock for 20 min to allow digestion of previously consumed bacterial food.
4. Prepare fresh homogenization buffer while worms are rocking. Keep buffer on ice.
5. Plate worms on NGM plates (lacking bacterial food) with glass spreader at room temperature (RT). Avoid areas of high concentration and overlapping worms.
6. Place the plates with worms into crosslinker and irradiate with UV-C (254 nm) at 0.3 J/cm^2 (*see* **Note 2**).
7. Collect worms in M9 buffer and resuspend in 1 mL of M9 and transfer to 2 mL tube.
8. Spin down for 30 s at 3000 rcf and resuspend the worms in 700 μL ice-cold homogenization buffer. Keep worms on ice from now on.
9. Sonicate the worms with 10 s pulses (5×) and 50 s rest periods on ice between sonications (~18 W RMS output power on the dismembrator).
10. Spin lysates at 16,000 × *g* for 15 min at 4 °C and collect supernatant.
11. Proceed to Subheading 3.2, or freeze lysates in dry-ice-ethanol bath for future use and store at −80 °C.

3.2 RNA Trimming and Immunoprecipitation

1. Aliquot 100 μL Protein G Dynabeads per sample.
2. Wash (2×) Protein G Dynabeads in homogenization buffer that does not contain the Proteinase Inhibitor Cocktail (PIC).
3. Resuspend beads in 100 μL homogenization buffer (w/o PIC).
4. Add 10 μg α-ALG-1 antibody.
5. Rock beads + antibody for 45 min at room temperature.
6. Wash beads + antibody (2×) with 900 μL homogenization buffer (w/o PIC).
7. Thaw lysates and quantify the concentration of protein content using Pierce BCA Protein Assay Kit.
8. Adjust lysate concentration to ~2–3 mg/mL with homogenization buffer with final volume of 1 mL (*see* **Note 3**).
9. Prepare 1:2000 dilution of RNase I in RNase I storage buffer.
10. Add 10 μL RNase I 1:2000 dilution and 2 μL Turbo DNase to 1 mL lysate (*see* **Note 4**).
11. Incubate samples for exactly 3 min in Thermomixer set to 37 °C and shaking at 1100 rpm.
12. Immediately transfer to ice.
13. Add 1 mL lysate to beads + antibody.
14. Rock beads + antibody + lysate at 4 °C for 1–2 h.
15. On ice, collect beads by placing tubes on magnetic rack for approximately 1 min, or until solution clears (*see* **Note 3**).
16. Discard supernatant.
17. Wash (2×) the beads with 900 μL ice-cold high-salt wash buffer (*see* **Note 5**).
18. Wash (2×) the beads with 900 μL ice-cold low-salt wash buffer.

3.3 Phosphorylation of Target RNA 5′ End

See Fig. 1 for ChimP "on-bead" biochemical steps.

1. Collect the beads with magnetic rack and remove supernatant.
2. Resuspend in PNK phosphorylation mix (20 μL per sample):

10× PNK buffer	2 μL
10 mM ATP	2 μL
T4 PNK 3′ phosphatase minus	1 μL
SUPERase in RNase inhibitor	0.5 μL
Nuclease-free water	14.5 μL

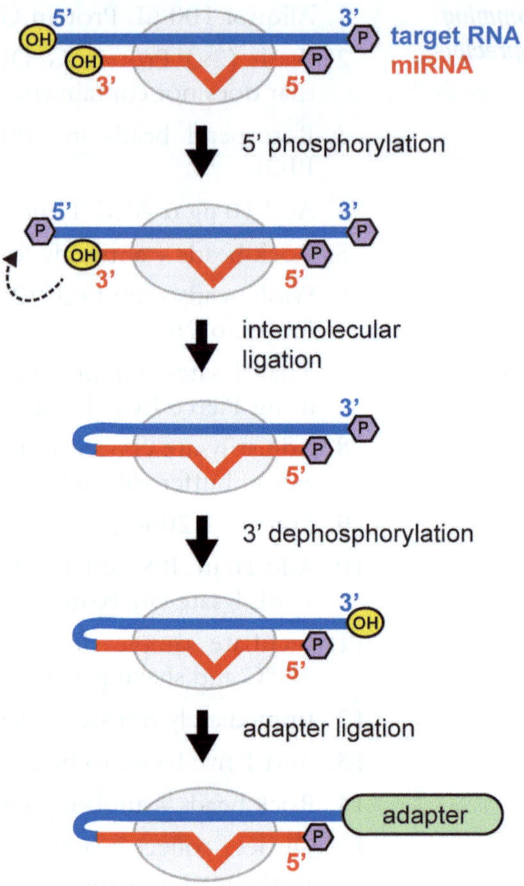

Fig. 1 Overview of the "on-bead" biochemical steps that generate chimeric RNAs and prepare the RNA for library amplification

3. Incubate at 37 °C for 10 min with intermittent shaking: 1100 rpm for 15 s every 5 min.
4. Wash (3×) with 900 μL low-salt wash buffer and leave in last wash on ice until ready to proceed.

3.4 Intermolecular Ligation of miRNA to Target RNA

1. Collect the beads with magnetic rack and remove supernatant.
2. Resuspend in intermolecular ligation mix (16 μL per sample):

Nuclease-free water	7.5 μL
4× RNA ligase buffer	5 μL
10 mM ATP	2 μL
T4 RNA ligase 1	1 μL
SUPERase in RNase inhibitor	0.5 μL

3. Add 4 μL of PEG400 to each reaction. Mix by gently pipetting up and down.
4. Incubate overnight at 16 °C with intermittent shaking: 1100 rpm for 15 s every 5 min.
5. Add 1 μL 10 mM ATP and 1 μL T4 RNA Ligase 1.
6. Incubate for 3 h at 16 °C with intermittent shaking: 1100 rpm for 15 s every 5 min.
7. Wash (1×) with 500 μL low-salt wash buffer.
8. Wash (1×) with 900 μL high-salt wash buffer.
9. Wash (2×) with 900 μL low-salt wash buffer and leave in second wash on ice until ready to proceed.

3.5 Removal of 3′ Phosphate

1. Collect the beads with magnetic rack and remove supernatant.
2. Resuspend in PNK 6.5 mix (20 μL per sample):

Nuclease-free water	15 μL
5× PNK pH 6.5 buffer	4 μL
T4 PNK enzyme	0.5 μL
SUPERase in RNase inhibitor	0.5 μL

3. Incubate at 37 °C for 20 min with intermittent shaking: 1100 rpm for 15 s every 5 min.
4. Wash (1×) with 500 μL low-salt wash buffer.
5. Wash (1×) with 900 μL high-salt wash buffer.
6. Wash (2×) with 900 μL low-salt wash buffer and leave in second wash on ice until ready to proceed.

3.6 3′ Linker Ligations

1. Collect the beads with magnetic rack and remove supernatant.
2. Resuspend in linker ligation mix (16 μL per sample):

Nuclease-free water	9 μL
4× ligation buffer	4 μL
T4 RNA ligase 1	1 μL
SUPERase in RNase inhibitor	0.5 μL
20 μM L3	1.5 μL

3. Add 4 μL of PEG400 to each reaction. Mix by gently pipetting up and down.

4. Incubate overnight at 16 °C with intermittent shaking: 1100 rpm for 15 s every 5 min.
5. Wash (1×) with 500 μL low-salt wash buffer.
6. Wash (2×) with 900 μL high-salt wash buffer.
7. Wash (2×) with 900 μL low-salt wash buffer and leave in second wash until ready to proceed.

3.7 Proteinase K Digestion and RNA Isolation

1. Prepare Proteinase K solution: 8 μL Proteinase K + 200 μL PK buffer per sample (*see* **Note 6**).
2. Incubate Proteinase K solution at 37 °C for 30 min to remove any RNases.
3. Wash beads (1×) in 100 μL PK Buffer.
4. Resuspend beads in 200 μL Proteinase K Solution.
5. Incubate at 37 °C at 1200 rpm for 45 min.
6. Add 200 μL of fresh 7 M urea +1× PK buffer and incubate for 30 min at 37 °C at 1000 rpm.
7. Add 500 μL of Phenol:Chloroform:Isoamyl Alcohol (25:24:1).
8. Incubate at 37 °C for 30 min at 1200 rpm.
9. Mix tubes vigorously by hand for a few seconds and centrifuge the tubes at 16,000 rcf for 5 min at room temperature.
10. Transfer the aqueous layer to a new microcentrifuge tube.
11. Add 400 μL Chloroform:Isoamyl Alcohol (24:1). Mix briefly and centrifuge at 16,000 rcf for 5 min at room temperature.
12. Transfer the aqueous layer to a new microcentrifuge tube and add:

3 M Sodium Acetate, pH 5.2	50 μL
Glycogen	1 μL
1:1 Ethanol:Isopropanol	1 mL

13. Precipitate overnight at −20 °C or −80 °C for 30 min.

3.8 Reverse Transcription

1. Spin precipitated RNA for 20 min at 16,000 rcf at 4 °C.
2. Remove the supernatant and wash the pellet (2×) with 500 μL 80% ethanol.
3. Air-dry pellet for exactly 8 min.
4. Resuspend the pellet in RNA/primer mix (7.25 μL per sample):

Nuclease-free water	6.25 μL
Rclip primer (0.5 pmol/μL)	0.5 μL
10 mM dNTP	0.5 μL

5. Incubate for 5 min at 70 °C.
6. Transfer to ice for 5 min.
7. Add reverse transcription mix (2.75 μL per sample) (*see* **Note 7**):

5× First-strand RT Buffer	2 μL
0.1 M DTT	0.5 μL
Superscript III	0.25 μL

8. On thermocycler incubate 5 min at 25 °C, 20 min at 42 °C, 40 min at 50 °C, and 5 min at 80 °C before cooling to 4 °C.

3.9 Size Selection of cDNA

1. Bring volume of reverse transcription reaction to 20 μL with nuclease-free water.
2. Add 36 μL (1.8× initial volume) AMPure XP beads.
3. Pipette entire volume (~50 μL) up and down 10 times to mix.
4. Incubate at room temperature for 10 min.
5. Place on magnetic rack for 5 min.
6. Remove 55 μL supernatant and discard.
7. Wash beads (2×) with 200 μL fresh 80% ethanol for 30 s at room temperature.
8. Air-dry beads for 5 min at room temperature.
9. Add 42.5 μL nuclease-free water to each tube and pipette up and down 10 times to mix.
10. Incubate for 2 min at room temperature.
11. Place onto magnetic rack for 5 min and transfer 40 μL supernatant to new tube.

3.10 Circularization and Linearization of cDNA

1. Add circularization mix (4.1 μL per sample) (*see* **Note 8**):

10× CircLigase Buffer II	2.2 μL
50 mM $MnCl_2$	1.1 μL
CircLigase II enzyme	0.8 μL

2. Incubate at 60 °C for 1 h, then 80 °C for 10 min in thermocycler.
3. Add 4 μL of 10× NEBuffer 3.1 and 1 μL cut_oligo_ddc.
4. In thermocycler, incubate for 1 min at 95 °C, then decrease temperature every 20 s by 1 °C until 25 °C has been reached.
5. Add 2 μL BamHI and mix by pipetting up and down gently.

6. Incubate for 30 min at 37 °C.
7. Add 30 μL of AmpureXP beads.
8. Pipette entire volume up and down 5 times to mix.
9. Add 75 μL of isopropanol.
10. Pipette entire volume up and down 5 times to mix.
11. Let stand at room temperature for 10 min.
12. Mix sample, and then let stand another 5 min at room temperature.
13. Place on magnetic rack for 5 min.
14. Remove supernatant and discard.
15. Wash beads (2×) with 200 μL fresh 80% ethanol for 30 s at room temperature.
16. Air-dry beads for 5 min at room temperature.
17. Add 20 μL nuclease-free water to each tube and pipette up and down 10 times to mix.
18. Heat for 2 min at 95 °C in thermocycler.
19. Immediately transfer to magnetic rack to collect beads.
20. Transfer supernatant to new PCR tube.

3.11 Library Amplification

1. Add PCR mix to each sample (21 μL per sample):

P5/P3Solexa Mix (10 μM each)	1 μL
Accuprime Supermix 1	20 μL

2. In thermocycler, run the following PCR program: 94 °C—2 min, [94 °C—15 s, 65 °C—30 s, 68 °C—30 s] $_{20\text{–}28\ \text{cycles}}$, 68 °C—3 min, 4 °C hold.
3. Mix 8 μL library with 2 μL 6× Loading Dye. Remaining library can be stored at −20 °C until ready to proceed.
4. Load library + dye on a precast 6% TBE gel with 1× TBE running buffer.
5. Run gel at 200 V for 30 min.
6. Stain gel with SYBR-GOLD for 10 min and image to confirm complex libraries and absence of residual P5_solexa and P3_solexa primers (*see* **Note 9**).

3.12 Chimera PCR

See Fig. 2 and **Note 10** for expected results from ChimP.

1. Prepare chimera target PCR samples and necessary controls (*see* **Note 11**):

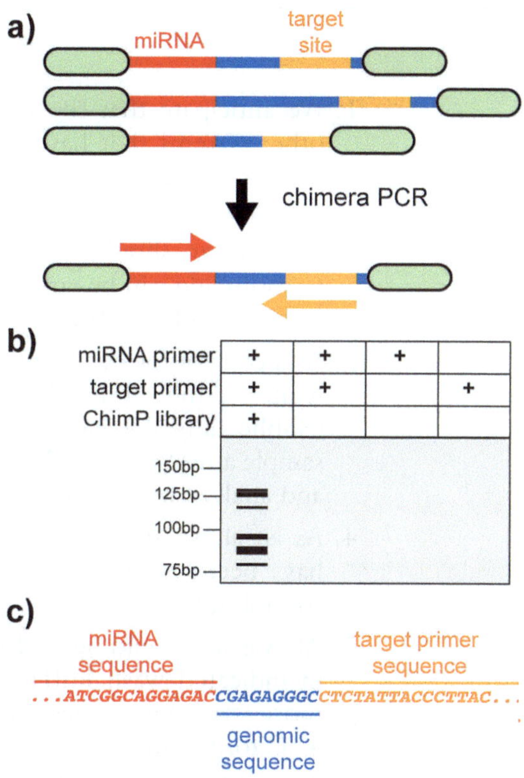

Fig. 2 (**a**) Diagram of the amplification of chimeric reads, which can be of various sizes due to different lengths of RNA before intermolecular ligation. Adapter sequences are shown in green, miRNA sequences are shown in red, miRNA target sites are shown in orange, and other RNA sequences are shown in blue. (**b**) Example of a ChimP experiment that includes no-template and single primer controls (*see Note 9*). (**c**) Sanger sequencing of ChimP products reveals genomic sequence not included in the target primer, which indicates a valid chimera

Phusion 2× HF Master Mix	10 μL
10 μM miRNA primer	0.5 μL
10 μM target reverse complement primer	0.5 μL
Library	0.5–2 μL
Nuclease-free library	To 20 μL

2. In thermocycler, run the following PCR program: 98 °C—2 min, [98 °C—15 s, 53 °C—30 s, 72 °C—30 s] $_{25–35\ \text{cycles}}$, 72 °C— 3 min, 4 °C hold (*see* **Note 12**).
3. Mix 8 μL library with 2 μL 6× Loading Dye.
4. Load ChimP product + dye on a precast 6% TBE gel with 1× TBE running buffer.
5. Run gel at 200 V for 30 min.
6. Stain the gel with SYBR-GOLD for 5 min and image (*see* **Note 13**).

4 Notes

1. We anticipate that the ChimP protocol will be adaptable to other systems that have a robust AGO antibody. The use of ChimP in other systems will require adjusting the homogenization buffer to one appropriate for that system.
2. *C. elegans* should be immobilized after UV treatment. Crosslinking can be enhanced by placing plates closer to bulbs or by using less M9 buffer.
3. As a control for the immunoprecipitation, save 15 μL of input lysate (Subheading 3.2, **step 8**), 15 μL of supernatant (Subheading 3.2, **step 14**), and 10% of immunoprecipitated sample after "on-bead" biochemistry (Subheading 3.6, **step 7**) and analyze by western blot.
4. As an alternative, performing RNase I digestions "on-bead" has been shown to increase reproducibility of RNA trimming [8].
5. All washes should be performed by gently resuspending beads in indicated wash buffer and inverting 5 times. If required, more stringent washes can be performed by rocking beads at 4 °C for 5 min.
6. Use of SDS instead of urea has been shown to increase release of RNA with proteinase K [8].
7. Use of TGRT-III and other reverse transcriptases may increase the amount of cDNA generated [8, 9].
8. Work from other groups has suggested that the CircLigase II reaction is inefficient [10]. We anticipate that using library generation steps that do not require CircLigase II will improve recovery of chimeric reads.
9. Excess P5/P3 solexa primers must be removed before carrying out ChimP PCRs. If libraries contain large amounts of primer-dimers, they can be cleaned up by either a final round of AMPure XP beads or gel purification.
10. We have tested ChimP extensively on the *C. elegans lin-41* 3′UTR and other let-7 target genes. For examples of ChimP gels, *see* ref. 5.
11. We recommend running single primer controls for both the miRNA primer and the target reverse complementary primer, as well as a no-template negative control. In addition to detecting off-target amplification, single primer controls will detect if excess P5/P3 solexa primers are leading to aberrant PCR products.
12. PCR annealing conditions should be determined based on the Tm of the miRNA forward primer. If the miRNA has a low

Tm, then the miRNA forward primer can be designed with some specificity for the library amplification primers.

13. Validation of ChimP products as valid chimeras is recommended when using new primer pairs. Validation can be performed by sequencing PCR products by sanger sequencing. Bona fide chimeras should include genomic sequences near the target primer sequence (Fig. 2c).

Acknowledgements

This work was supported by an NSF Graduate Research Fellowship to J. P. B. (DGE-1144086) and a grant from the NIH (GM071654) to A. E. P.

References

1. Jonas S, Izaurralde E (2015) Towards a molecular understanding of microRNA-mediated gene silencing. Nat Rev Genet 16:421–433. https://doi.org/10.1038/nrg3965
2. Pasquinelli AE (2012) MicroRNAs and their targets: recognition, regulation and an emerging reciprocal relationship. Nat Rev Genet 13:271–282. https://doi.org/10.1038/nrg3162
3. Helwak A, Kudla G, Dudnakova T, Tollervey D (2013) Mapping the human miRNA interactome by CLASH reveals frequent noncanonical binding. Cell 153:654–665. https://doi.org/10.1016/j.cell.2013.03.043
4. Grosswendt S, Filipchyk A, Manzano M et al (2014) Unambiguous identification of miRNA: target site interactions by different types of ligation reactions. Mol Cell 54:1042–1054. https://doi.org/10.1016/j.molcel.2014.03.049
5. Broughton JP, Lovci MT, Huang JL et al (2016) Pairing beyond the seed supports MicroRNA targeting specificity. Mol Cell 64:320–333. https://doi.org/10.1016/j.molcel.2016.09.004
6. Moore MJ, Scheel TKH, Luna JM et al (2015) miRNA–target chimeras reveal miRNA 3′-end pairing as a major determinant of Argonaute target specificity. Nat Commun 6:8864. https://doi.org/10.1038/ncomms9864
7. König J, Zarnack K, Luscombe NM, Ule J (2012) Protein-RNA interactions: new genomic technologies and perspectives. Nat Rev Genet 13:77–83. https://doi.org/10.1038/nrg3141
8. Zarnegar BJ, Flynn RA, Shen Y et al (2016) irCLIP platform for efficient characterization of protein-RNA interactions. Nat Methods 13:489–492. https://doi.org/10.1038/nmeth.3840
9. Van Nostrand EL, Shishkin AA, Pratt GA et al (2017) Variation in single-nucleotide sensitivity of eCLIP derived from reverse transcription conditions. Methods 126:29–37. https://doi.org/10.1016/j.ymeth.2017.08.002
10. Van Nostrand EL, Pratt GA, Shishkin AA et al (2016) Robust transcriptome-wide discovery of RNA-binding protein binding sites with enhanced CLIP (eCLIP). Nat Methods 13:508–514. https://doi.org/10.1038/nmeth.3810

Chapter 13

High-Quality Overlapping Paired-End Reads for the Detection of A-to-I Editing on Small RNA

Josephine Galipon, Rintaro Ishii, Soh Ishiguro, Yutaka Suzuki, Shinji Kondo, Mariko Okada-Hatakeyama, Masaru Tomita, and Kumiko Ui-Tei

Abstract

Paired-end RNA sequencing (RNA-seq) is usually applied to the quantification of long transcripts such as messenger or long non-coding RNAs, in which case overlapping pairs are discarded. In contrast, RNA-seq on short RNAs (≤ 200 nt) is typically carried out in single-end mode, as the additional cost associated with paired-end would only translate into redundant sequence information. Here, we exploit paired-end sequencing of short RNAs as a strategy to filter out sequencing errors and apply this method to the identification of adenosine-to-inosine (A-to-I) RNA editing events on human precursor microRNA (pre-miRNA) and mature miRNA. Combined with RNA immunoprecipitation sequencing (RIP-seq) of A-to-I RNA editing enzymes, this method takes full advantage of deep sequencing technology to identify RNA editing sites with unprecedented resolution in terms of editing efficiency.

Key words RNA-seq, Small RNA, Paired-end sequencing, RNA editing, Inosine, microRNA

1 Introduction

Illumina RNA sequencing (RNA-seq) is sufficient for the relative quantification of RNA, but its error rate remains problematic when the desired application is the detection of single nucleotide modifications on RNA, due to the difficulty of distinguishing between true RNA editing events and sequencing errors. Because of this, Sanger direct sequencing remains the gold standard for validating novel editing sites. In the case of adenosine deaminases acting on RNA (ADAR)-mediated adenosine-to-inosine (A-to-I) editing sites, only sites with an editing ratio above about 5% to 10%—meaning that more than 10% of transcripts are edited at that

Josephine Galipon and Rintaro Ishii contributed equally to this work.

Ulf Andersson Vang Ørom (ed.), *miRNA Biogenesis: Methods and Protocols*, Methods in Molecular Biology, vol. 1823, https://doi.org/10.1007/978-1-4939-8624-8_13, © Springer Science+Business Media, LLC, part of Springer Nature 2018

position—may be clearly distinguished from a typical Sanger sequencing background [1]. For this reason, the vast majority of RNA editing studies establish an arbitrary cut-off of 5~10% for the detection of editing sites, making it extremely difficult to publish reports of lower efficiency sites. Nevertheless, editing sites with low editing ratios have the potential to be biologically important. For instance, in the case of extremely abundant transcripts such as the oncogenic miR-21, editing levels as low as 0.1~1.0% have the potential to generate a subpopulation of edited miR-21 that is abundant enough to target and downregulate a entirely novel set of genes. Indeed, editing in the miRNA seed region responsible for target recognition leads to genome-wide retargeting of the edited microRNA (miRNA) [2, 3]. For example, in order to validate an editing site with a ratio of 0.1~1.0%, in theory one would need to read 2000~20,000 individual clones by Sanger sequencing to obtain just two reads with an A-to-G substitution corresponding to that editing site. This is currently unrealistic in terms of time and cost, although it may be possible to overcome this by simultaneous overexpression of both the ADAR and the candidate pre-miRNA in an attempt to artificially bump up the editing ratio. In conclusion, if the accuracy problem of RNA-seq is overcome, deep sequencing is expected to reveal more of these low-efficiency editing sites with potentially important biological functions.

Here, we used paired-end sequencing on small RNAs as a trick to filter out most errors, decreasing the theoretical error frequency to about 1 in 10^6 bases. We applied this method to the detection of A-to-I editing sites on human precursor and mature miRNA by RNA immunoprecipitation of ADAR-bound small RNA. The paired-end sequencing method reads each cDNA fragment from both the 5′- and the 3′-end, and is therefore not an essential requirement for the relative quantification of small RNAs, as the reverse and forward end of the pair would frequently overlap. However, we reasoned that since the 5′- and 3′-ends are each read by an independent sequencing reaction, paired-end sequencing may be used as an internal validation to distinguish sequencing errors from genuine RNA editing sites, by eliminating the paired-end reads for which the overlapping forward and reverse sequences do not perfectly match. Candidate editing sites are then statistically validated by the log-likelihood ratio (LLR) test using the Phred quality scores produced by both the forward and the reverse end, in an approach adapted from previous literature [4–6].

To implement this method, we aimed to identify and compare the precursor and mature miRNA editing target preferences of two constitutive nuclear isoforms of ADARs: ADAR1-p110 and ADAR2. To this end, we overexpressed each isoform separately as mycGFP (mGFP)-tagged constructs as described previously [7]. Two popular methods for investigating RNA–protein interactions in vivo are RNA immunoprecipitation (RIP) and cross-linking immunoprecipitation (CLIP) or photoactivatable-ribonucleoside-enhanced crosslinking (PAR-CLIP), a method relying on the incorporation of photoreactive ribonucleoside analogs, were successfully used to identify genome-wide miRNA–protein interactions [8, 9]. However, aside from difficulties in identifying the precise location of A-to-I editing site due to crosslinking-induced mutations, CLIP-seq and its variants may introduce a bias in favor of imperfect double-stranded RNA (dsRNA) binding targets, as unpaired nucleotides are more available for UV crosslink formation due to better accessibility of the A-form helix major groove [10]. Previously, Methylene blue (MB) as a photoreactive intercalating agent was successfully used to induce crosslinking of dsRNA to protein by visible light [11], but MB mediates heavy oxidative damage to purines [12]. Crosslinking methods are therefore inadequate for the precise downstream identification of RNA editing sites. Out of these concerns, we opted for straightforward RNA immunoprecipitation sequencing (RIP-seq) of overexpressed mycGFP-tagged ADAR isoforms with mycGFP as a control to detect background editing by endogenous ADARs [7].

To capture both pre- and mature miRNAs, high-throughput strand-specific sequencing of small RNA (sized 15–110 nucleotides) was carried out in 101 nucleotide paired-end mode. The reads produced were then mapped to the human genome using a method for identifying A-to-I editing sites known as *collapse mapping* [13], which we adapted to strand-specific paired-end data. After generation of high-quality super-reads from the overlap between both sides of a pair, statistical validation by the LLR test [4–6], and discarding of sites that overlapped with annotated single nucleotide polymorphisms (SNPs) [14–18], allowed the identification of 31 unique editing sites on mature and pre-miRNA with editing ratios, 12 of which were confirmed by other studies and presented here ([2, 19–27, Table 1). Among these 12 canonical sites, 5 had editing ratios below 5%, ranging from 0.21 to 4.76%, suggesting that our method successfully identifies relatively low-efficiency editing sites.

2 Materials

2.1 Recommendations for Handling RNA

The following general precautions should be followed at all times to avoid RNA degradation. If these steps are respected, it is not necessary to prepare buffers and reagents with DEPC-treated water.

1. Work on ice and use gloves; while wearing the gloves avoid contact with things that are regularly in contact with human skin (such as door knobs or garbage bin lids).
2. Use a set of pipettes that is specific for RNA work and have never been used for handling cell cultures; if unavailable, the use of filter pipette tips is recommended.

Table 1
Statistically validated candidate A-to-I editing sites that overlapped with previous studies

Name	Chr	Str	GRCh37/hg19	Pos	A + G	G	ER (%)	LLR	Sample
pre-let-7g (5p)	3	–	52302364	10	6	6	100.000	6.6	ADAR1 IP
pre-mir-21 (3p)	17	+	57918679	8	674	2	0.297	2.7	ADAR1 input
pre-mir-140 (5p)	16	+	69967021	16	14	2	14.286	5.0	ADAR2 IP
pre-mir-455 (5p)	9	+	116971745	17	10	4	40.000	13.0	ADAR1 input
pre-mir-1304 (3p)	11	–	93466874	5	18	8	44.444	7.1	ADAR1 IP
let-7d-3p	9	+	96941181	5	186	10	5.376	25.5	ADAR2 input
miR-22-3p	**17**	**–**	**1617215**	**15**	**23978**	**62**	**0.259**	**59.1**	**ADAR1 IP**
miR-22-3p	**17**	**–**	**1617215**	**15**	**13992**	**30**	**0.214**	**11.6**	**ADAR2 IP**
miR-22-3p	**17**	**–**	**1617215**	**15**	**105860**	**548**	**0.518**	**35.2**	**ADAR2 input**
miR-24-2-5p	19	–	13947156	6	84	4	4.762	4.4	ADAR1 IP
miR-151a-3p	8	–	141742704	3	290	2	0.690	10.4	ADAR1 input
miR-210-5p	11	–	568122	12	486	2	0.412	2.3	ADAR2 input
miR-340-3p	5	–	179442328	13	34	2	5.882	4.4	ADAR2 input
miR-1301-3p	2	–	25551539	5	24	2	8.333	5.1	ADAR2 input

Name: name of the miRNA precursor or mature miRNA on which the candidate editing site was identified. For precursors, only reads longer than 25 nt were taken into account; reads that were 24 nt or shorter were considered as coming from mature miRNAs, and the position of the editing site relative to the mature sequences was indicated as (5p) or (3p) in the case of miRNA precursors; *Chr* chromosome; *Str* strand; *hg19* position on the corresponding chromosome of the human genome UCSC hg19 release. *Pos* position relative to the +1 nucleotide of the corresponding mature miRNA. A + G: total coverage of A's and G's at the editing site. G: coverage of G's at the editing site. ER (%): editing ratio. *LLR* log-likelihood ratio calculated as described in Subheading 3.7. Sample: sample in which the editing site was detected

3. Use commercial RNase-free water for resuspension, dilution, and enzymatic reaction steps involving RNA.
4. All buffers and reagents should be 0.22 μm filter-sterilized, as autoclaving would release the RNases contained in micro-organic contaminants, an undetermined portion of which may renature after the solution cools down.
5. Samples should be harvested quickly with a method that disrupts cells and inactivates nucleases.

2.2 Cell Culture and Transfection

1. 9-cm cell culture dishes.
2. HeLa cells or cell line of interest.
3. 37 °C CO_2 incubator.
4. Cell culture medium: Dulbecco's modified Eagle Medium (DMEM), 4.5 g/L glucose, without antibiotics, supplemented with sodium pyruvate and 10% heat-inactivated fetal bovine serum (FBS).

5. Opti-MEM™ I Reduced Serum Medium.
6. Lipofectamine™ 2000 Transfection Reagent.
7. pcDNA3.1 or equivalent expression vector containing the coding region for the tagged protein of interest, in this case mycGFP-ADAR1-p150, -p110 and mycGFP-ADAR2 (not covered here). For the method to clone the gene of interest into the expression vector for mycGFP-ADAR isoform cloning, please refer to [7].
8. DAPI (4′,6-diamidino-2-phenylindole) staining reagent.

2.3 RNA Immunoprecipitation and Deep Sequencing (RIP-Seq)

1. An antibody against the tag of interest, in this case rabbit polyclonal anti-GFP antibody (antigen: Full-Length *Aequorea victoria* GFP).
2. Goat anti-rabbit immunoglobulin G (IgG) conjugated with alkaline phosphatase.
3. 1× PBS(−): 137 mM NaCl, 2.7 mM KCl, 10 mM Na_2HPO_4, 1.8 mM KH_2PO_4 (pH 7.4 adjusted with HCl).
4. RiboCluster Profiler™ RIP-Assay Kit.
5. TURBO™ DNase (2 U/μL).
6. SuperSep™ Ace 5–20% gradient pre-casted polyacrylamide gel.
7. BioTrace™ PVDF membrane.
8. Tris/Glycine/SDS electrophoresis buffer: 0.3% (w/v) Tris base, 1.44 (w/v) glycine, 0.1% (w/v) SDS, pH 8.3 (no pH adjustment required).
9. TBS-T buffer: 20 mM Tris-HCl (pH 7.5), 150 mM NaCl, 0.2% Triton X-100 (for washing, supplemented with 5% skim milk for blocking).
10. Alkaline phosphatase buffer: 100 mM Tris-HCl (pH 9.5), 100 mM NaCl, 5 mM $MgCl_2$.
11. CDP-Star® Detection Reagent.
12. RiboZero Kit.
13. Illumina® TruSeq Small RNA Library Prep Kit Reference Guide.

2.4 Hardware and Equipment

1. Bio-Rad Trans-Blot® Turbo™ Transfer System, iBlot® Dry Blotting System, or equivalent.
2. NanoDrop™ 2000 Spectrophotometer, QuBit Fluorometer, or equivalent.
3. Thermocycler for polymerase chain reaction (PCR).
4. Fluorescence microscope with filters compatible with GFP and DAPI.
5. Illumina® HiSeq 2000 Sequencing System.

2.5 Software and Database Versions

Although the method can be implemented with later versions without any major problems, the versions that were used for this study are as follows:

1. CASAVA v1.8.2 (provided by Illumina).
2. Human genome sequence: Genome Reference Consortium GRCh37, UCSC human hg19.
3. Alignment software: bowtie 0.12.7 [28].
4. Non-coding RNA database: NONCODE v3.0 [29].
5. Repetitive sequences: RepeatMasker3.3.0 (http://www.repeatmasker.org).
6. microRNA database: miRBase v19 [30].
7. SNP databases: 1000 genomes [14], dbSNP [15], HapMap [16], miRNASNP [17], and a publication from Han et al. [18].

3 Methods

3.1 Cell Culture and Transfection

Pre-warm all cell media and 1× PBS unless otherwise specified. Note that the cell dishes for fluorescence imaging involving DAPI staining of the nucleus were separate from those for RNA immunoprecipitation, and that antibiotics were not used.

1. Plate 3×10^6 HeLa cells per 9-cm cell culture dish in cell culture medium and culture overnight in a CO_2 incubator at 37 °C.
2. Rinse in pre warmed 1× PBS and change medium to DMEM without FBS.
3. Prepare transfection mix following instructions for Lipofectamine™ 2000, but using 5.5 μg of tagged protein expression vector per 9-cm dish in Opti-MEM™ I Reduced Serum Medium.
4. Add the transfection mix to each cell culture.
5. After incubating 4 h in a 37 °C CO_2 incubator, remove the transfection media and replace by cell culture medium.
6. Incubate in a CO_2 incubator at 37 °C for 2 days.
7. To observe cell confluency and GFP-tagged protein expression by fluorescence microscopy, remove culture media and add 1× PBS(−).
8. Observe localization of mycGFP-tagged proteins by DAPI staining of nuclear DNA (*see* Fig. 1a) (*see* **Note 1**).
9. For RNA immunoprecipitation, proceed to Subheading 3.2.

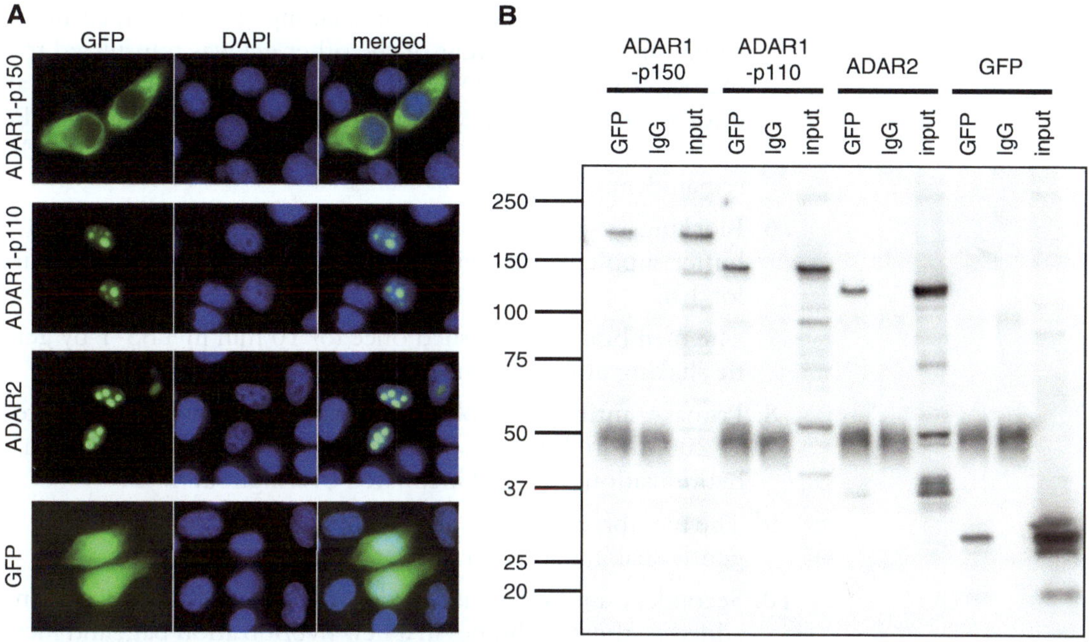

Fig. 1 *Subcellular localization of full-length mycGFP-ADAR proteins.* (**a**) Representative images of fluorescence microscopy of HeLa cells 48 h after transfection. GFP signals were merged with DAPI staining of nuclear DNA. This data shows that mycGFP-ADAR-p150 is mostly in the cytoplasm, whereas mycGFP-ADAR1-p110 and mycGFP-ADAR2 localize in the nucleus, concentrated in nucleolar foci. Control mycGFP expression is ubiquitous. (**b**) Western blot using anti-GFP antibody confirms expression and successful immunoprecipitation of full-length mycGFP-ADAR proteins. From left to right, GFP: immunoprecipitation using an anti-GFP antibody, IgG: control immunoprecipitation using goat anti-rabbit immunoglobulin G (IgG), input: total cell lysate before immunoprecipitation. The expected lengths for mycGFP-ADAR1-p150, mycGFP-ADAR1-p110, mycGFP-ADAR2, and mycGFP are 165 kDa, 133 kDa, 110 kDA, and 29 kDA, respectively. A background signal observed at around 50 kDa is consistent with the size of the antibody heavy chain and is not present in the total cell lysate. Immunoprecipitation resulted in a unique band at the expected size for all four isoforms, which also corresponded exactly to the most intense band in the respective total cell lysate. This means that overexpression and immunoprecipitation of ADAR isoforms were successful

3.2 RNA Immunoprecipitation, Silver Staining, and Western Blot

1. Wash HeLa cells 5 times in pre-warmed 1× PBS(−).
2. Cell lysis and RNA immunoprecipitation (IP) was carried out following the maker's instructions for the RiboCluster Profiler™ RIP-Assay Kit. Cell lysates were immunoprecipitated using 15 ng of either rabbit anti-GFP antibody or control IgG.
3. 2 μL of total cell lysate and 100 μL of the final elution mixture for anti-GFP and control IgG IP samples were set aside for silver staining and Western blotting before proceeding to RNA extraction.
4. Two identical 5–20% gradient polyacrylamide gels were purchased for SDS-PAGE electrophoresis of the samples obtained in Step 3 in Tris/Glycine/SDS electrophoresis buffer at 20 mA for 80 min.

5. One gel may be used for silver staining (not described here), while the protein content of the other gel was transferred to a methanol-washed PVDF membrane by electroblotting in an iBlot® Dry Transfer System with default parameters (P0 7:00 min: 20 V for 1 min, 23 V for 4 min, 25 V for the remainder).
6. Blocking of the PVDF membrane was performed in TBS-T buffer supplemented with 5% skim milk at room temperature for 1 h.
7. The membrane was washed once for 10 min in TBS-T by gentle shaking at room temperature.
8. Primary antibody (rabbit polyclonal anti-GFP) was diluted 1/2000th in 1 mL of TBS-T, added to the membrane in a hybridization bag, and incubated overnight at 4 °C.
9. The membrane was washed 3 times 10 min in TBS-T buffer by gentle shaking at room temperature.
10. Secondary antibody was pre-diluted 1/10,000th in 1 mL and added to the membranes in a new hybridization bag, and incubated at 4 °C overnight.
11. The membrane was washed 3 times 10 min in TBS-T buffer by gentle shaking at room temperature.
12. Lay the membrane onto a piece of Saran wrap, and cover it with alkaline phosphatase (AP) buffer.
13. Discard AP buffer, cover with CDP-Star Detection Reagent, and proceed to imaging (*see* Fig. 1b).
14. The RNA extracted from input and IP samples at Step 3 was kept for small RNA sequencing in Subheading 3.3.

3.3 Library Preparation and Strand-Specific Paired-End Deep Sequencing

1. The concentration of RNA extracted in Subheading 3.2. for input and IP samples was measured using a NanoDrop™ 2000 spectrophotometer or Qubit fluorometer.
2. DNA was digested using 0.88U of TURBO™ DNase.
3. Ribosomal RNA was removed using the RiboZero™ Magnetic Gold Kit (*see* **Note 2**).
4. Strand-specific libraries for deep sequencing were prepared according to instructions in the Illumina® TruSeq Small RNA Library Prep Kit Reference Guide, and the amplified cDNA templates were size-selected on a polyacrylamide gel to obtain fragments that correspond to cDNAs that would be 15–110 nucleotides long without the adapters. This length range is expected to recover most mature and precursor miRNAs (pre-miRNAs).
5. 101 nucleotide paired-end sequencing with HiSeq 2000 yielded between 14,539,145 and 25,450,999 paired-end reads

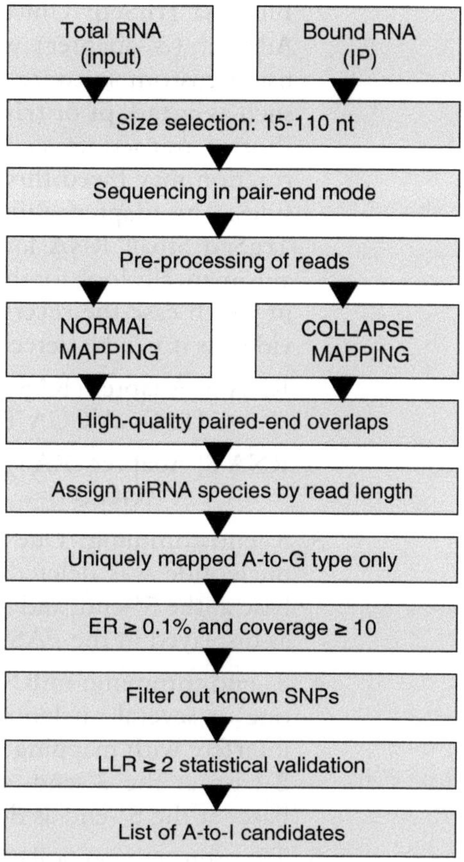

Fig. 2 *Bioinformatics pipeline for paired-end sequencing-based identification of A-to-I editing sites on small RNA*. This figure summarizes the steps described in Subheadings 3.4–3.7

depending on the sample. Given the read length of 101 nucleotides, and the fact that cDNAs ranging 15–110 nucleotides were size-selected, the full-length mature miRNA and at least part of the pre-miRNA regions are expected to be read twice, once in each direction.

3.4 Data Pre-Processing

An overview of the bioinformatics workflow from Subheadings 3.4–3.7 is available (*see* Fig. 2).

1. **CASAVA** v1.8.2 was used to generate **FASTQ** files in Illumina 1.8+/Phred+33 format, for which raw read quality values (QV) range between 0 and 41. These FASTQ files contain a label for each read specifying whether a read has "passed filtering" or not, indicated by Y or N, with Y meaning that the read failed the test. Therefore, only reads labeled with an N were kept for further analysis.

2. Illumina TruSeq® Universal Adapter (5′-adapter) and Indexed Adapter (3′-adapter) were trimmed. This may be achieved using custom software or tools that handle paired-end reads, such as **cutadapt** or **trimmomatic**. For reads originating from shorter transcripts such as mature miRNA, the sequencing reaction may bleed through into the adapter regions on both sides. The adapter sequences were as follows, according to the TruSeq Small RNA kit; please note that most tools do not automatically look for the reverse complement of the 5′-adapter, in which case the reverse complement of RA5 should be provided as it will be detected on the reverse reads:

   RNA 5′ Adapter (RA5) 5′-GTT CAG AGT TCT ACA GTC CGA CGA TC-3′

   RNA 3′ Adapter (RA3) 3′-TGG AAT TCT CGG GTG CCA AGG-3′

3. 5′-end trimming: Due to relatively poor quality, the 5′-most nucleotide was deleted by trimming the forward reads by 1 base at the 5′-end, and the reverse reads by 1 base at the 3′-end as displayed in the **FASTQ** file.
4. 3′-end trimming: miRNAs are often subject to the addition of one or several nucleotides at their 3′-end [31], which could interfere with mapping. Thus, forward reads were trimmed by 2 base at the 3′-end, and reverse reads were trimmed by 2 bases at the 5′-end as displayed in the **FASTQ** file.
5. To ensure a theoretical error of less than 1 in 1000 bases (10^{-3}) for all positions on the read, the read sequences were scanned from the 5′-end and any downstream sequences were trimmed upon encountering a base with a quality value (QV) lower than 30.
6. Reads that were less than 10 bases long following the above steps were discarded. Depending on the sample, 33~65% of reads remained.

3.5 Collapse Mapping and Calling of RNA Editing Sites

Most mapping algorithms are not optimal for detecting mismatches in long reads, due to limitations in the number of allowed mismatches. Here, we chose to adopt a method called *collapse mapping* that enables the mapping of edited reads regardless of the number or distribution of a given mismatch type within the read [13], and adapted it as below to make it compatible with strand-specific paired-end data.

1. First, an index is generated for the human genome (GRCh37, UCSC human hg19) using **bowtie-index**, which comes with **bowtie** (we used version 0.12.7 at the time, but any other version may be employed) [28] (*see* **Note 3**).
2. ***Normal mapping***: **bowtie** was used to map reads with *unique hits* and *no mismatches* to the human genome with help from the index generated in Step 1.

3. The overlapping region was extracted from pairs for which the forward and reverse overlapped without any mismatches; here we obtained between 0.9 and 1.8 million high-quality *super-reads* while also retaining strand information.
4. ***Collapse mapping***: **bowtie** mapping was performed again but this time allowing *multiple hits* and *no mismatches*, and the unmapped reads were output to a separate file.
5. The unmapped reads from Step 3 were then collapsed to a three-base system: all A's were converted to G's, on both the forward and the reverse reads. These reads, that now contain only 3 types of bases, are referred to as *collapsed reads*.
6. Two collapsed versions of the human genome (+) strand were prepared: one that was collapsed from A to G, and the other from T to C (*see* Fig. 3a). These are hereby referred to as *collapsed genomes*.

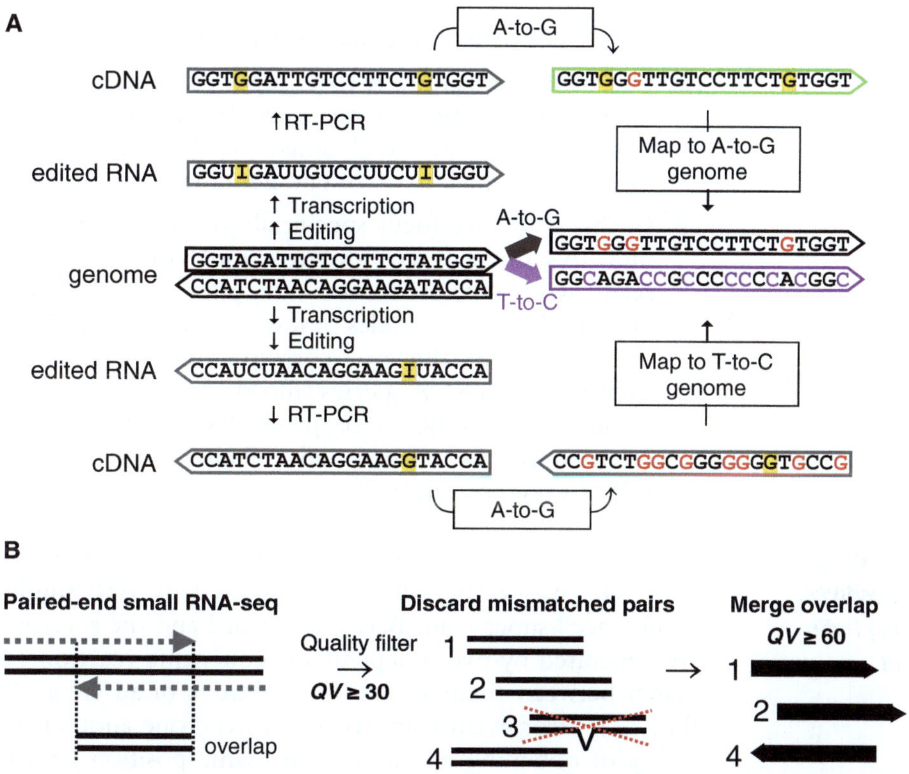

Fig. 3 *Collapse mapping of paired-end reads and generation of high-quality super-reads from the overlapping sequences of paired-ends.* (**a**) Upper half: A-to-G collapsed forward reads are mapped to a three-base A-to-G collapsed (+) strand genome. Lower half: A-to-G collapsed reverse reads are mapped to a T-to-C collapsed (+) strand genome. (**b**) Left: the overlap between two reads of a pair is retrieved. Middle: overlaps that contain mismatches are discarded to filter out sequencing errors. Right: the overlaps' forward and reverse strand are merged into one high-quality sequence while conserving strand-specific information. These images were modified from the *Handbook on RNA-Seq Experiments*, Chapter 2, Part III [32] with permission from Yodosha, Co., Ltd.

7. Using **bowtie**, the collapsed forward and reverse reads were mapped to the A-to-G and T-to-C collapsed genomes, respectively; *no mismatches* were allowed, and reads that mapped to more than 20 locations are discarded.
8. Reverse and forward reads of a same pair that fail to overlap on the same genomic location are discarded, and the overlapping region was extracted from the remaining pairs in our case.
9. Paired-end reads for which the overlapping sequences do not perfectly match, meaning that it contains one or more mismatches between the forward and the reverse, are discarded, as this is most likely indicative of a sequencing error (*see* Fig. 3b); in our case, this eliminated 0.02~0.05% of pairs.
10. The overlapping part of the paired-end read was reverted to its original 4-base sequence, and compared to the sequence of the original 4-base genome to which it mapped. Among these super-reads, those containing mismatches relative to the genome other than the "A-to-G" type were discarded. Here, 12~17% of reads contained "A-to-G" type only.
11. Out of the remaining A-to-G super-reads, those that mapped to multiple positions in the genome (around 80%) were discarded. In the end, we obtained 74,113 to 290,561 collapsed super-reads which may contain genuine editing sites.
12. In this study, we focus specifically on reads that map to pre-miRNA sequences retrieved from miRBase (version v19) [29]; reads were separated into categories according to length: "mature" (24 nucleotides or less) or "non-mature" (25 nucleotides or more), the latter corresponding for the most part to non-mature miRNA species (mostly pre-miRNA, and possibly some primary miRNA or pri-miRNA). The distribution of read lengths for normal and collapse reads that mapped to miRNA regions can be found (*see* Fig. 4).

3.6 Probability of False Positives, and Editing Ratio Threshold

In the previous section, the overlapping sequence of each overlapping pair was retrieved and considered as one single "high-confidence" super-read. Because forward and the reverse of a pair are generated by two independent sequencing reactions, any mismatch between them should be indicative of an error. Assuming that each error event is independent from one another, the probability of obtaining an error at the same position i on both the forward and the reverse is the product of their probability of error, defined by the QV of the base encountered at position i. In the present scenario, only bases with a QV of 30 or above are retained, so that the probability of double error occurrence should never go above 10^{-6} (1 in 1,000,000 bases) (*see* **Note 4**). Furthermore, the probability of the same error type occurring on both strands at the same position should be even lower than this. It ensues that the

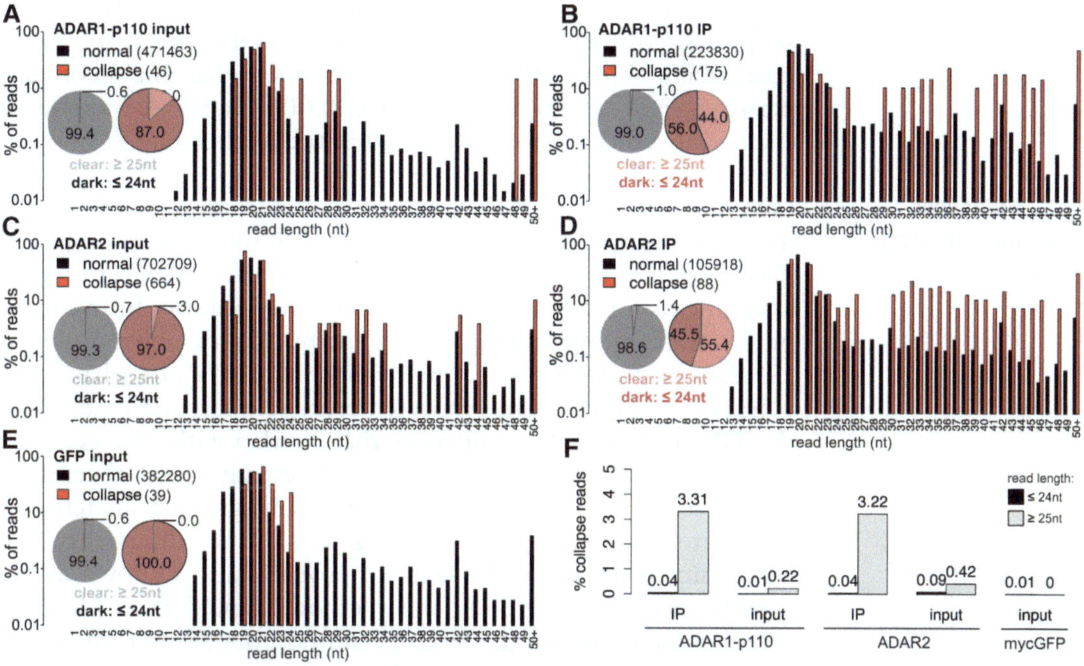

Fig. 4 *Distribution of read lengths for reads that mapped to annotated miRNA regions.* (**a–e**) Read length distribution of reads mapped to miRNA regions by collapse (red) and normal (dark gray) mappings of input (**a**, **c**, **e**) and IP (**b**, **d**) samples from the cells expressing mycGFP-ADAR1-p110 (ADAR1-p110; **a**, **b**), mycGFP-ADAR2 (ADAR2; **c**, **d**), and mycGFP control (GFP; **e**), respectively; *X*-axis: read length in nt, *Y*-axis: percentage of read number shown in the $\log_{10}$ scale. Pie charts summarize the respective percentages of longer reads assumed to originate from miRNA precursors ($\geq$ 25 nt) and mature miRNA ($\leq$ 24 nt) in the collapse and normal mappings of input and IP samples. (**f**) Percentage of collapse reads in longer ($\geq$ 25 nt) and shorter reads ($\leq$ 24 nt) for each sample, showing a clear enrichment in precursor miRNA-like collapse reads in the IP fractions of mycGFP-ADAR1-p110 and mycGFP-ADAR2. This may be indicative of binding and editing by ADARs on the double-stranded regions of pre-miRNA stem loops

probability of a false positive is negligible. For instance, in order for such a false positive to count as a candidate RNA editing site, an A-to-G error on the forward end should correspond to the complementary T-to-C error at the exact same genomic position on the overlapping reverse end of the paired-end read.

Given this information, the editing ratio (ER) of a candidate editing site, defined as the percentage of reads at a harboring an A-to-G type mismatch at a given genomic position, which means A-to-I RNA editing, is given by the following formula which uses the coverage of each nucleotide type at that particular position:

$$\mathrm{ER} = \left([G]_{\mathrm{collapse}} + [G]_{\mathrm{normal}}\right) / \left([A,G]_{\mathrm{collapse}} + [A,G]_{\mathrm{normal}}\right) \times 100 \quad (1)$$

Sites with an ER $\geq$ 0.1% and a total coverage of 10 or more (5 × 2 ends of a paired-end read) were kept as candidate A-to-I editing sites for statistical validation in Subheading 3.7. It should be noted that the threshold defined as ER ≥ 0.1% (1 in 1000 transcripts edited at a particular position) ensures that this A-to-G type mismatch is at least 1000 times more frequent than the theoretical per-read error occurrence (1 in 1,000,000 reads). A method to statistically validate this is presented in the Subheading 3.7.

3.7 Statistical Validation of Editing Sites

Over the past decade, there has been much debate about the validity of deep sequencing-based identification of RNA editing sites. False positives may arise both from sequencing errors, from the assignment of reads to the wrong genomic location during mapping, and from overlapping with single nucleotide polymorphisms (SNPs) (*see* **Note 5**). Since the quality value (QV) of a base provides an indication of the expected error rate for that base, using the QV of that base on all reads that overlap with that genomic position, it is possible to calculate the average local theoretical error rate, and therefore the probability that a given mismatch is an error rate, versus the probability that it is a true editing site. The logarithm of the ratio of these two probabilities is known as the log-likelihood ratio (LLR). This method adapted from [4–6] to accommodate the fact that each position in our super-read is read twice independently:

1. Given n reads mapping to a given position i, if x of reads contain a mismatch at position i, we assume that x follows the binomial distribution:

$$\mathrm{f}\left(p;n,x,i\right) = {}_{n}\mathrm{C}_{x} \times \mathrm{p}^{x} \times \left(1-\mathrm{p}\right)^{(n-x)} \quad (2)$$

2. The maximum likelihood p of the binomial distribution in [1] is then:

$$p = x / n \quad (3)$$

3. To account for sequencing errors in overlapping paired-end reads, the observed error rate ε is defined by the sum Φ of the two quality values at position i originating from both ends of the paired-end read, such as $\varepsilon = 10^{-\Phi}$. From there, the maximum likelihood estimator r taking ε into account can be defined as:

$$r = \left\{\Sigma\left[i\epsilon\ G\right]\left(1-\varepsilon_i\right) + \Sigma\left[i\epsilon\ A\right]\left(\varepsilon_i\right)\right\} / n \quad (4)$$

4. Then, the maximum likelihood $\max_r$ defined in [3] is compared to the probability $P(n|r = 0, \varepsilon, i)$ of all reads containing a sequencing error at position i (in which case, $r = 0$) using the following log-likelihood ratio (LLR):

$$\mathrm{LLR} = \log_{10}\left\{\max_r\left[P\left(n|,,r|,,\varepsilon|,,i\right)\right] / P(n|,,r = 0|,,\varepsilon|,,i)\right\} \quad (5)$$

5. Finally, we selected the candidate editing sites for which LLR ≥ 2, meaning that the mismatch at position i is 100 times more likely to be a true editing site than a sequencing error, given two quality values per position i from the paired-end sequencing.

4 Notes

1. The editing ratio is strongly affected by the relative level of expression of ADAR protein and their target miRNAs, respectively. Therefore, transfection efficiency of the expression construct is expected to influence the editing ratio. This paired-end sequencing-based strategy is useful for the identification of editing sites with a wide range of editing ratios, including low-efficiency sites.
2. Epicentre's RiboZero™ Gold Machetic Kit contains enough reagents for six reactions, but we were successful in using the reagents to a dilution of 6/7 in order to accommodate seven samples.
3. Bowtie 1 (bowtie) and Bowtie 2 (bowtie2) are two very different implementations, and their indexes are not interchangeable. While bowtie2 is generally the better choice when aligning reads that are longer than 50 bp, bowtie is sometimes faster and/or more sensitive when aligning short reads, but ignores unpaired alignments. Even when 101 nt in paired-end mode, reads originating from mature miRNA reads are very short (18–24 bases). Furthermore, trimming steps and merging of paired-end overlaps generates reads for pre-miRNA that are shorter than the initial read length generated by the sequencer (peak between 40 and 50 bases). For these reasons, we recommend Bowtie 1, but the choice is up to the reader.
4. Sequence quality filtering results in a trade-off between quality and coverage. In this method, we prioritized quality over coverage, but the thresholds suggested here may be tuned to optimize coverage. Indeed, if the general quality of the sequencing data is lower, a full-length QV ≥ 30 cut-off may filter out most of the reads. For instance, one may decide on a QV cut-off of 25 (1 error every $10^{2.5}$ or 316 bases), resulting in a maximum error on the paired-end overlapping regions of 10^{-5} (1 error every 100,000 bases). In this case, one may consider editing sites with an ER $\geq$ 1% as valid candidates, or keep it at ER $\geq$ 0.1% and decide on more stringent coverage and/or LLR test thresholds.
5. The possibility remains that true editing sites overlap with known SNPs. Similarly, the opposite case where novel stable SNPs, present only in this cell line and absent from public

databases, were falsely interpreted as editing sites. However, it is reasonable to expect that in this case, the editing ratio would be either very close to 50% in the case of a heterozygote allele, or very close to 100% in case of a homozygote. Here, one editing site on pre-let-7 g had an editing ratio of 100%. Nonetheless, this site was confirmed by several other independent studies [2, 22, 24] and is conserved in mouse [22], suggesting that this is a genuine editing site. Combining the paired-end sequencing trick presented here with DNA-seq to compare the genome and transcriptome of the cell line of interest will be ideal for the future experimental setup of RNA editing studies.

Acknowledgments

This work was supported by grants from the Ministry of Education, Culture, Sports, Science and Technology (MEXT) of Japan to KU-T. Keio University Institute for Advanced Bioscience affiliates were supported by research funds from the Yamagata prefectural government and the City of Tsuruoka.

References

1. Ihle MA, Fassunke J, König K, Grünewald I, Schlaak M, Kreuzberg N, Tietze L, Schildhaus HU, Büttner R, Merkelbach-Bruse S (2014) Comparison of high resolution melting analysis, pyrosequencing, next generation sequencing and immunohistochemistry to conventional sanger sequencing for the detection of p.V600E and non-p.V600E BRAF mutations. BMC Cancer 14:13
2. Kawahara Y, Zinshteyn B, Sethupathy P, Iizasa H, Hatzigeorgiou AG, Nishikura K (2007) Redirection of silencing targets by adenosine-to-inosine editing of miRNAs. Science 315(5815):1137–1140
3. Kume H, Hino K, Galipon J, Ui-Tei K (2014) A-to-I editing in the miRNA seed region regulates target mRNA selection and silencing efficiency. Nucleic Acids Res 42:10050–10060
4. Li JB*, Levanon EY*, Yoon JK, Aach J, Xie B, LeProust E, Zhang K, Gao Y, Church GM (2009) Genome-wide identification of human RNA editing sites by massively parallel DNA capturing and sequencing. Science 324:1210–1213 *equal contribution
5. Bahn JH, Lee JH, Li G, Greer C, Peng G, Xiao X (2012) Accurate identification of A-to-I RNA editing in human by transcriptome sequencing. Genome Res 22:142–150
6. Chepelev I (2012) Detection of RNA editing events in human cells using high-throughput sequencing. Methods Mol Biol 815:91–102
7. Galipon J, Ishii R, Suzuki Y, Tomita M, Ui-Tei K (2017) Differential binding of three major human ADAR isoforms to coding and long non-coding transcripts. Genes (Basel) 8(2):68
8. Hafner M, Landthaler M, Burger L, Khorshid M, Hausser J, Berninger P, Rothballer A, Ascano M, Jungkamp AC, Munschauer M, Ulrich A, Wardle GS, Scott Dewell S, Zavolan Z, Tuschl T (2010) J Vis Exp 41:2034–2039
9. Chi SW, Zang JB, Mele A, Darnell RB (2009) Argonaute HITS-CLIP decodes microRNA-mRNA interaction maps. Nature 460:479–486
10. Weeks KM, Crothers DM (1993) Major groove accessibility of RNA. Science 261:1574–1577
11. Liu ZR, Sargueil B, Smith CW (2000) Methylene blue-mediated cross-linking of proteins to double-stranded RNA. Methods Enzymol 318:22–33
12. Tuite EM, Kelly JM (1993) Photochemical interactions of methylene blue and analogues with DNA and other biological substrates. J Photochem Photobiol B 21:103–124
13. Wu D, Lamm AT, Fire AZ (2011) Competition between ADAR and RNAi pathways for an

extensive class of RNA targets. Nat Struct Mol Biol 18:1094–1101

14. 1000 Genomes Project Consortium, Abecasis GR, Auton A, Brooks LD, MA DP, Durbin RM, Handsaker RE, Kang HM, Marth GT, McVean GA (2012) An integrated map of genetic variation from 1,092 human genomes. Nature 491:56–65
15. Sherry ST, Ward MH, Kholodov M, Baker J, Phan L, Smigielski EM, Sirotkin K (2001) dbSNP: the NCBI database of genetic variation. Nucleic Acids Res 29:308–311
16. The International HapMap Consortium (2003) The international HapMap project. Nature 426:789–796
17. Gong J, Liu C, Liu W, Wu Y, Ma Z, Chen H, Guo AY 2015 An update of miRNASNP database for better SNP selection by GWAS data, miRNA expression and online tools. Database (Oxford) 2015:bav029
18. Han M, Zheng Y (2013) Comprehensive analysis of single nucleotide polymorphisms in human microRNAs. PLoS One 8:e78028
19. Kawahara Y, Megraw M, Kreider E, Iizasa H, Valente L, Hatzigeorgiou AG, Nishikura K (2008) Frequency and fate of microRNA editing in human brain. Nucleic Acids Res 36:5270–5280
20. Luciano DJ, Mirsky H, Vendetti NJ, Maas S (2004) RNA editing of a miRNA precursor. RNA 10:1174–1177
21. García-López J, Hourcade Jde D, Del Mazo J (2013) Reprogramming of microRNAs by adenosine-to-inosine editing and the selective elimination of edited microRNA precursors in mouse oocytes and preimplantation embryos. Nucleic Acids Res 41:5483–5493
22. Ekdahl Y, Farahani HS, Behm M, Lagergren J, Öhman M (2012) A-to-I editing of microRNAs in the mammalian brain increases during development. Genome Res 22:1477–1487
23. Alon S, Mor E, Vigneault F, Church GM, Locatelli F, Galeano F, Gallo A, Shomron N, Eisenberg E (2012) Systematic identification of edited microRNAs in the human brain. Genome Res 22:1533–1540
24. Ramaswami G, Li JB (2014) RADAR: a rigorously annotated database of A-to-I RNA editing. Nucleic Acids Res 42(Database issue):D109–D113
25. Warnefors M, Liechti A, Halbert J, Valloton D, Kaessmann H (2014) Conserved microRNA editing in mammalian evolution, development and disease. Genome Biol 15:R83
26. Wang Y, Xu X, Yu S, Jeong KJ, Zhou Z, Han L, Tsang YH, Li J, Chen H, Mangala LS, Yuan Y, Eterovic AK, Lu Y, Sood AK, Scott KL, Mills GB, Liang H (2017) Systematic characterization of A-to-I RNA editing hotspots in microRNAs across human cancers. Genome Res 27(7):1112-1125
27. Tomaselli S, Galeano F, Alon S, Raho S, Galardi S, Polito V, Presutti C, Vincenti S, Eisenberg E, Locatelli F, Gallo A (2015) Modulation of microRNA editing, expression and processing by ADAR2 deaminase in glioblastoma. Genome Biol 16(1):5
28. Langmead B, Trapnell C, Pop M, Salzberg SL (2009) Ultrafast and memory-efficient alignment of short DNA sequences to the human genome. Genome Biol 10:R25
29. Bu D, Yu K, Sun S, Xie C, Skogerbø G, Miao R, Xiao H, Liao Q, Luo H, Zhao G, Zhao H, Liu Z, Liu C, Chen R, Zhao Y (2012) NONCODE v3.0: integrative annotation of long noncoding RNAs. Nucleic Acids Res 40:D210–D215
30. Kozomara A, Griffiths-Jones S (2011) miRBase: integrating microRNA annotation and deep-sequencing data. Nucleic Acids Res 39:D152–D157
31. Neilson JR, Sandberg R (2010) Heterogeneity in mammalian RNA 3′ end formation. Exp Cell Res 316:1357–1364
32. Galipon J, Ishiguro S, Tomita M, Ui-Tei K (2016) Detection of A-to-I RNA editing by RNA-Seq. NGS application, Handbook of RNA-Seq experiment, 113–117, Yodosha.

Chapter 14

Targeting miRNA for Therapeutics Using a Micronome Based Method for Identification of miRNA-mRNA Pairs and Validation of Key Regulator miRNA

Parveen Bansal, Ashish Kumar, Sudhir Chandna, Malika Arora, and Renu Bansal

Abstract

MicroRNAs are 18–22 bp long non-coding sequences and play a critical role in diverse biological processes, through modulation of gene expression at the post-transcriptional level by binding at the 3′-untranslated region of target mRNA. Consequent upon the discovery of structural and functional features of miRNA targeting, several molecular methods have been developed to identify miRNA targets. However, these methods suffer several drawbacks, including technical challenges, requirement of high cell volumes, inability to differentiate between direct and indirect targets, cell/tissue as well as experimental-specificity and imprecise binding site information. Alternatively in silico approach enables the exploration of the potential miRNA-mRNA pairs to investigate signature miRNA and proteins involved in the signaling of various diseases. Here, we describe micronome-based standard method for identification of miRNA-mRNA pairs as well as validation of key regulator miRNA.

Key words MicroRNA, Search Tool for Retrieval of Interacting Genes/Proteins (STRING), miRNA Target, Protein–protein Interaction, RNase protection assay, mirTARbase

1 Introduction

MicroRNAs play a vital role in regulating various cellular and metabolic pathways by controlling cell proliferation, differentiation, survival, and cell death [1]. Studying miRNAs–mRNA interaction is emerging as a central tool as well as a promising field for generating novel RNA-based treatment and diagnostic modalities for a plethora of dreaded diseases including cancer [2]. miRNAs are achieving a lot of attention and recognition due to its involvement in gene expression; however, the quantum and extent of translational control exerted by miRNAs is unknown. In addition, it also has been observed that expression of miRNA may alter in various diseases in contrast to normal individuals during their biogenesis due to dysregulation of various genes. In case of altered expressions,

Ulf Andersson Vang Ørom (ed.), *miRNA Biogenesis: Methods and Protocols*, Methods in Molecular Biology, vol. 1823,
https://doi.org/10.1007/978-1-4939-8624-8_14,

some of the miRNA are overexpressed and start acting as oncogenes known as "oncomirs" whereas due to reduced expression, these may act as tumor suppressors [3]. Approximately two thousands of miRNA are identified in human and this small number controls around 30,000 of the genes, but currently, only a handful of specific targets have been experimentally validated.

Consequent upon the discovery of new structural and functional features of miRNA targeting, a huge amount of their mRNA target expression data has been generated with use of high-throughput technologies such as microarrays and Next Generation Sequencing (NGS) [4–6]. The data generated is so complex that it would be unrealistic to use it for intended purposes. To overcome this, a large number of online bioinformatics tools are available free of cost to manage such huge data [7–10]. Both basic and applied miRNA research are being revolutionized by computational tools and databases. Screening of such a huge number of miRNAs targets in experimental studies is tedious, expensive, and time consuming, hence there is a need to specifically identify and validate the target miRNA molecules using bioinformatics tool that can provide fast solutions. Standard operative procedure for the same is needed. Here authors provide a micronome-based standard method for identification of miRNA-mRNA pairs which have high potential to be targeted for diagnostic and treatment applications in oral squamous cell carcinoma in present case [11, 12].

2 Materials

2.1 Databases Required

1. **STRING v10:** STRING database has stored information compiled from various sources like experimental data, computational prediction methods as well as data stored in public text collections. This database is freely accessible and available online. Moreover it is updated on the regular basis. Latest version of this software contains information regarding 200 organisms and around 9.6 million of proteins. This database was used for gene ontology of the selected genes.
2. **miRTarBase:** This database has stored information and more than three hundred and sixty thousand miRNA-target interactions (MTIs) which has been collected by manual surveying of pertinent literature as well as by filtering data presented in various research articles related to functional studies of miRNAs. This database provides the most updated information as compared to other databases. This database was used for retrieval of miRNA of desired genes.
3. **Cytoscape 3.3.0:** Cytoscape is a bioinformatics-based software available as online database which is freely accessible and acts as a platform for visualizing molecular interactions among

genes, integrating with the gene expression profiles, observing the interaction networks as well as other associated data. In addition to these features, additional features can be extended in the form of plugins. These plugins are available for network as well as molecular profiling, to get insight into new layouts, additional file format support and connection with databases and searching in vast networks. This database was used for filtering the gene targets of the selected genes for miRNA network visualization.

3 Methods

3.1 Identification and Screening of Differentially Expressed Genes in Relation to Gene of Interest

1. Consider all the differentially expressed genes for analysis from already available databases related to gene of interest. As herein, differentially expressed genes were considered for analysis from oral cancer databases.
2. Maintain a record of all the identified differentially expressed genes in a master Microsoft excel sheet [*see* **Note 1**].
3. The scope of inclusion of database may include the data available in published articles, on Pubmed or other reputed database websites (Web of science, Scopus, google scholar, etc.) articles related to functional concept like herein inflammation, apoptosis, metastasis, angiogenesis, etc. For example, a total of four databases were selected for carrying out the identification.

3.2 Gene Ontology to Identify Genes

1. Carry out gene ontology (i.e., representation of gene and gene product attributes) using STRING v10 [13], a most commonly used bioinformatics tool available free of cost online. Gene ontology is needed to be carried out to gain further insight into the biological processes potentially mediated by desired genes.
2. The genes may be categorized and targeted as per the requirements of the study. Here we had categorized the genes involved in inflammation and analyzed the functional categories of genes by performing gene ontology (GO) analysis.
3. Here we have targeted various biological processes related to inflammatory response, positive regulation of inflammatory response, and negative regulation of inflammatory response.

3.3 Retrieval of miRNA of Target Genes

1. Search the involvement of the genes for their respective regulating miRNA.
2. Arrange all the regulatory genes and miRNA in separate excel sheets [*see* **Note 2**].

3. Retrieve miRNA of the genes of interest under study using miRTarBase database. miRTarBase is required to perform retrieval of the miRNA of desired genes as it represents experimentally validated miRNA-target interactions (MTIs) from research articles and is more frequently updated than other databases [14].
4. Retrieve positive and negative regulators separately from miRTarBase. e.g., TarBase [15], miRecords [16], and miR2Disease [17].
5. Herein the retrieval of miRNA of genes obtained for inflammation was done. miRTarBase revealed 1574 MTIs for the 75 genes involved in OSCC inflammation. Similarly, 12 genes with 318 MTIs were found to be involved in negative regulation, whereas 19 genes with 385 MTIs were found to be involved in positive regulation of inflammation.

3.4 Computational Method of miRNA Target Filtering

1. Initially scan the direct mRNA–miRNA interactions using miRNA prediction databases. The miRNA prediction databases are available for searching direct targets for miRNAs.
2. miRNA with good miSVR scores and binding at conserved regions should be selected to further identify the universal miRNA. For the filtration part, deregulated genes with their targeted miRNAs should be arranged in such a file which can be opened in Cytoscape 3.3.0 [18] for miRNA network visualization (Micronome).
3. Herein, microrna.org and TargetScan [19] visualization (Micronome) databases were used for mRNA-miRNA alignment and further filtration of miRNA extracted from miRTarBase [*see* **Note 2**].

3.5 Construction of Regulation Networks of miRNA-mRNA (Micronome)

1. Maintain all the files obtained from microrna.org and TargetScan.org in the same format to get an insight on the pathways and regulatory genes available in the micronome.
2. Based on the miRNA, build the micronome using Cytoscape 3.3.0.
3. Examine the interaction between the set of associated genes at the miRNA level on the basis of number of nodes in the network, the total number of interactions in the network, and number of nodes with degree greater than three [20].

Herein, authors present the construction of three micronomes using above mentioned database, i.e., Cytoscape version 3.3.0. This database has represented miRNAs involved in inflammatory response along with its positive and negative regulation. As per the database study, it has been revealed that in case of negative regulation 173 miRNAs have 311 interactions with 53 target genes

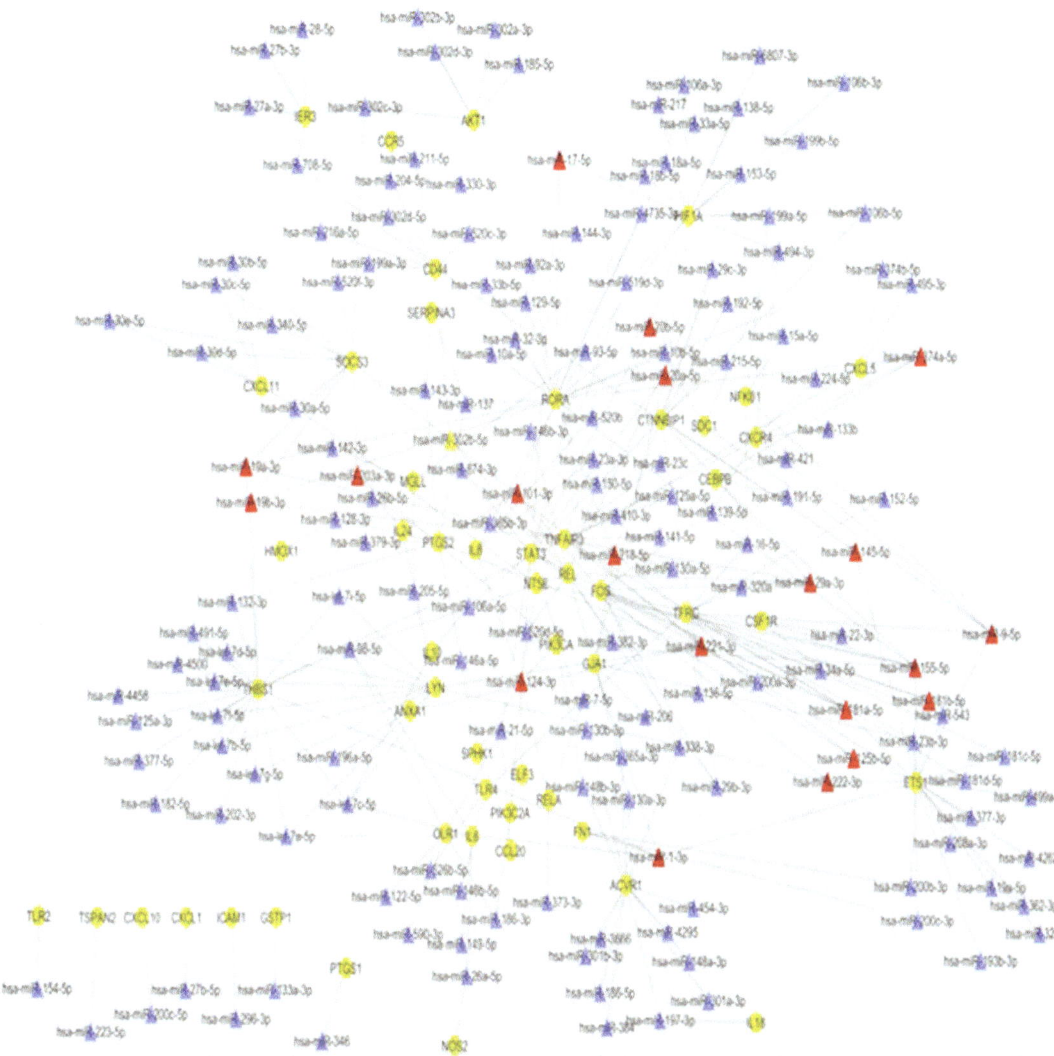

Fig. 1 Representation of miRNA–mRNA interaction network of inflammatory response and negative regulation of inflammatory response demonstrating 173 validated miRNAs having 311 interactions with 53 target genes in inflammation. Purple triangle nodes represent miRNAs while red triangle nodes represent common miRNAs of both processes. Yellow circles are the genes of inflammatory response and its negative regulation [reproduced from ref. 12]

(Fig. 1), whereas in case of positive regulation it showed 160 miRNAs having 280 interactions with 55 target genes (Fig. 2).

3.6 Protein–Protein Interaction [PPI] Network

1. Perform PPI analysis in order to visualize the association of key genes with the genes under study using STRING v10. PPI is required for elucidating signaling pathway. Signaling PPI analysis is required to be performed in order to discuss the association of key genes involved in inflammation.

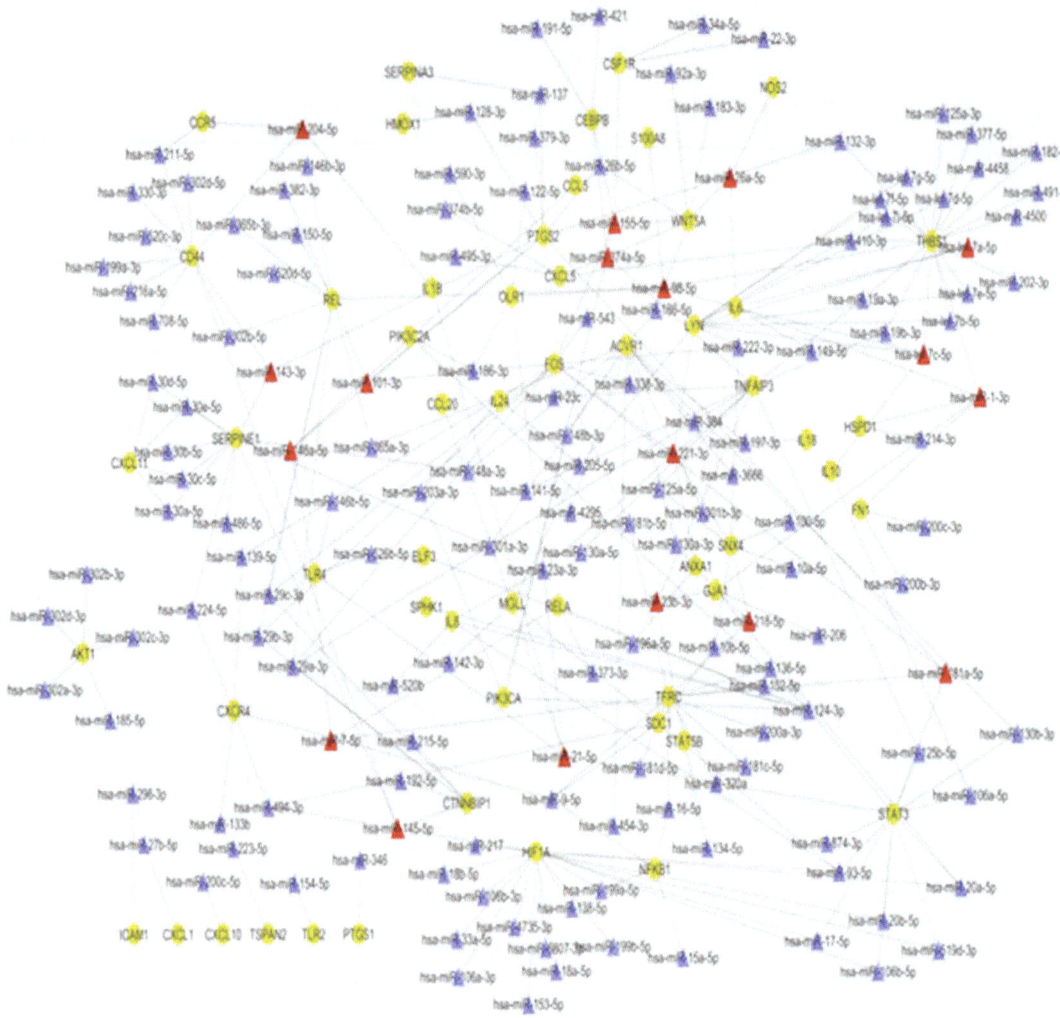

Fig. 2 miRNA–mRNA interaction network of inflammatory response and positive regulation of inflammatory response demonstrating 160 validated miRNAs having 280 interactions with 55 target genes in inflammation. Purple triangle nodes represent miRNAs while red triangle nodes represent common miRNAs of both processes. Yellow circles are the genes of inflammatory response and its positive regulation [reproduced from Ref. 12]

2. Reveal the key gene involved in the regulation using STRING database [*see* **Note 3**].

Herein, we had constructed a final micronome to demonstrate the presence of universal miRNA among three processes, i.e., inflammation, positive regulation of inflammation, and negative regulation of inflammation. From the data, it was observed that eight universal miRNAs were involved in regulation of inflammation in OSCC (Fig. 3) [11]. In addition to construction of a micronome, interaction network showing miRNA–mRNA interactions was generated for both negative regulation and positive regulation

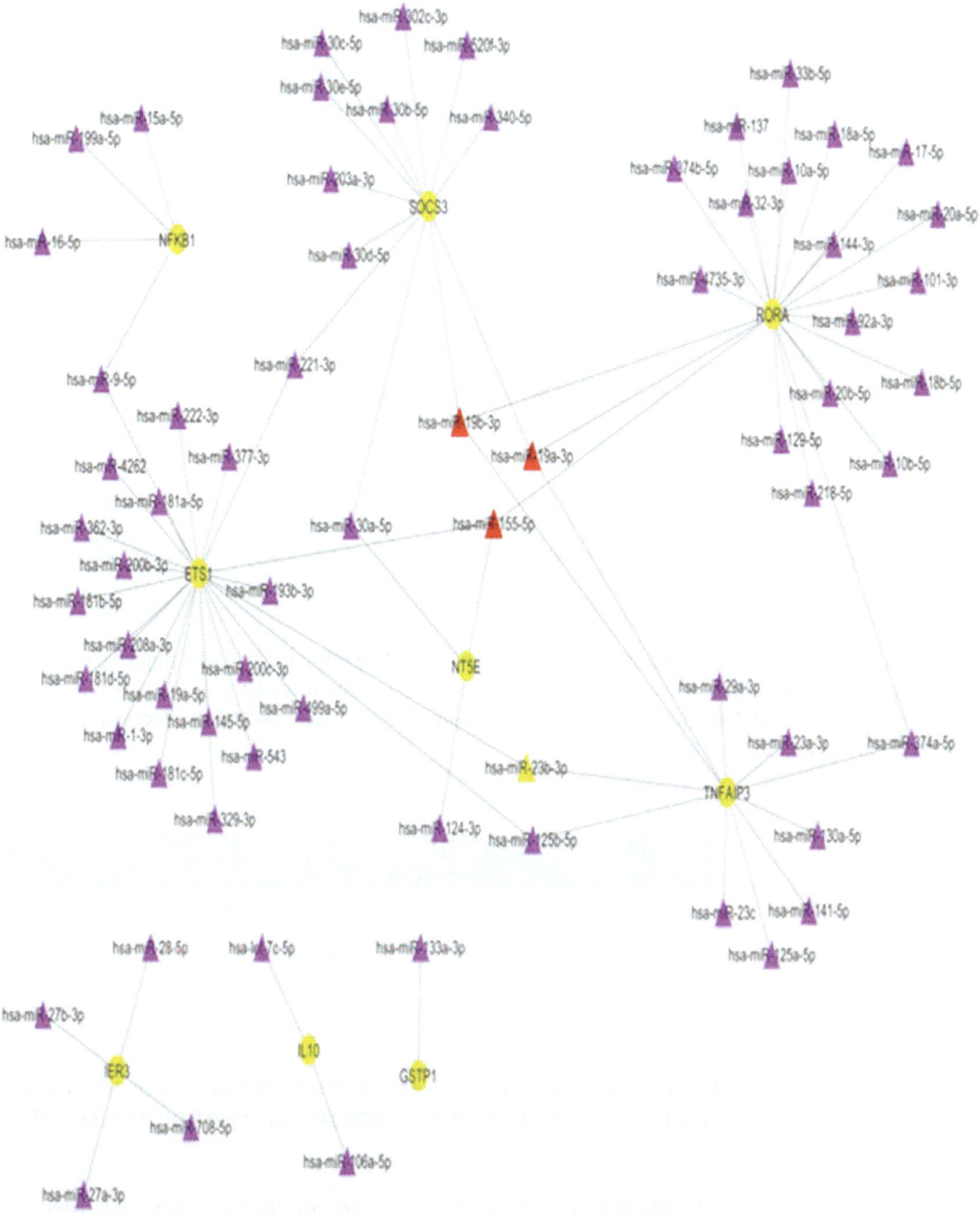

Fig. 3 Negative regulation of inflammatory response network between miRNA and genes. Purple triangle nodes indicate miRNAs while red color triangles indicate key miRNAs (hsa-miR-19a-3p, hsa-mir-19b-3p, and hsa-mir-155-5p) in the process. Yellow represents gene nodes [reproduced from ref. 12]

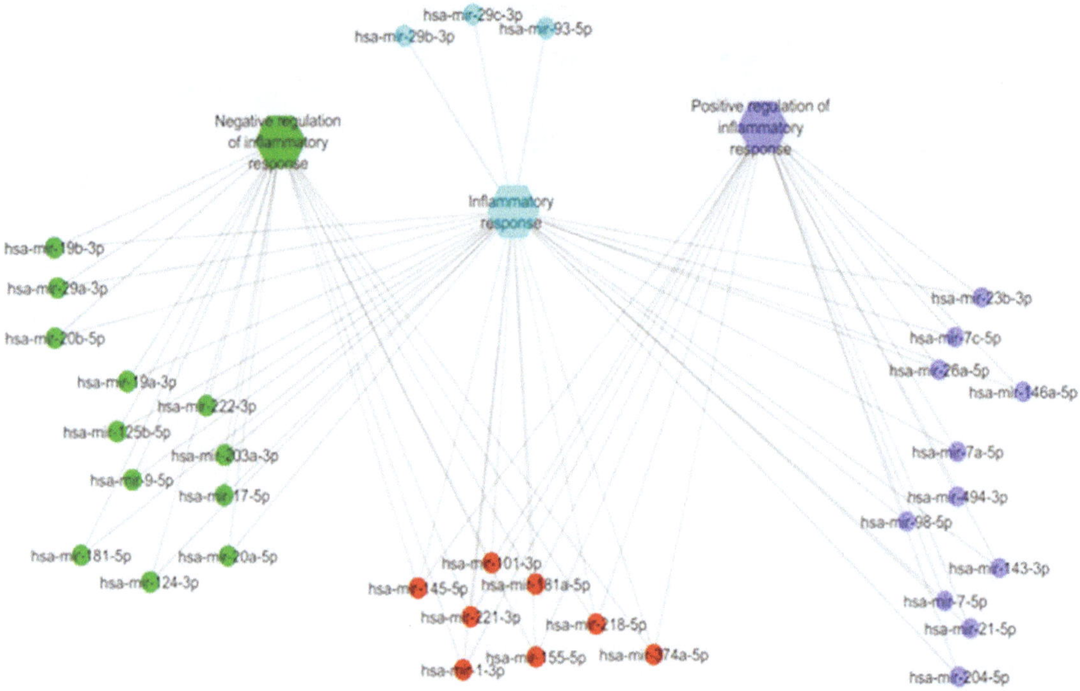

Fig. 4 Final micronome with universal-miRNAs targeting inflammatory response, negative regulation and positive regulation of inflammatory response. The micronome revealed eight universal miRNAs (hsa-mir-155-5p, hsa-mir-103-3p, hsa-mir-145-5p, hsa-mir-181-5p, hsa-mir-221-3p, hsa-mir-218-5p, hsa-mir-1-3p, and hsa-mir-374a-5p) [reproduced from ref. 12]

		Predicted consequential pairing of target region (top) and miRNA (bottom)
Position 1561-1568 of SOCS3 3' UTR	5'	...CAAUCCAAGUCAAACUUUGCACA...
hsa-miR-19a-3p	3'	AGUCAAAACGUAUCUAAACGUGU
Position 1561-1568 of SOCS3 3' UTR	5'	...CAAUCCAAGUCAAACUUUGCACA...
hsa-miR-19b-3p	3'	AGUCAAAACGUACCUAAACGUGU

Fig. 5 Sequence alignment between hsa-mir-19a/b and its direct target SOCS3 (as shown by prediction database TargetScan.org) [reproduced from ref. 12]

of inflammatory response. Herein, interaction network revealed the miR-19a, miR-19b, and miR-155 as key regulating miRNA as shown in Fig. 4. Moreover while visualizing the interactions, SOCS3, RORA, and TNFAIP3 were found to be direct target of miR-19a and miR-19. RORA and ETS1 were supposed to be the direct targets of miR-19a/19b, whereas NT5E was the direct target of mir-155-3p as shown in Figs. 4 and 5, respectively. Apart from this, it is pertinent to mention that no significant information

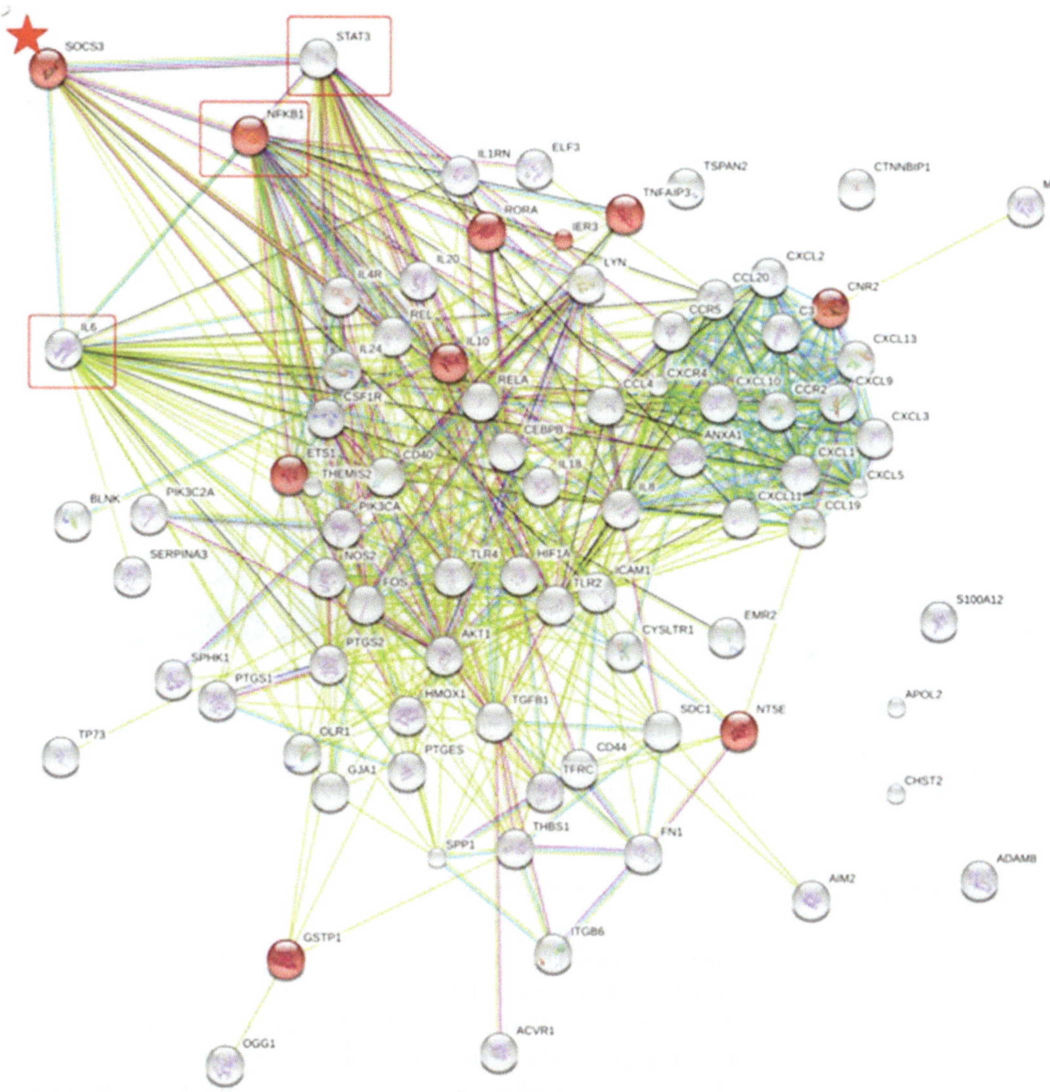

Fig. 6 PPI network of gene of inflammatory response and negative regulation (PPI enrichment p-value zero). Negative regulator gene SOCS3 (target of hsa-mir-19a) is directly connected with NF-κB-IL-6-STAT3. Genes SOCS3, IL-6, and STAT3 are the key members of JAK-STAT signaling pathway (inflammation in OSCC). The NF-κB-IL-6-STAT3 cascade is critical link in cancer inflammation as NF-κB induces the expression of inflammatory cytokines and anti-apoptotic genes, key enzymes in prostaglandin synthase pathway and angiogenic factors; STAT3 is a point of convergence of numerous oncogenic signaling pathways; IL-6 promotes the proliferation of tumor-initiating cells and possesses anti-apoptotic activity [reproduced from ref. 12]

was detected in positive regulation of inflammatory response. PPI network of inflammatory protein revealed SOCS3 as key regulator gene as it has been found that this gene was strongly connected with the transcription factor (NF-kB), interleukin cytokine (IL-6), and signal transducer activator, i.e., STAT3, as shown in Fig. 6.

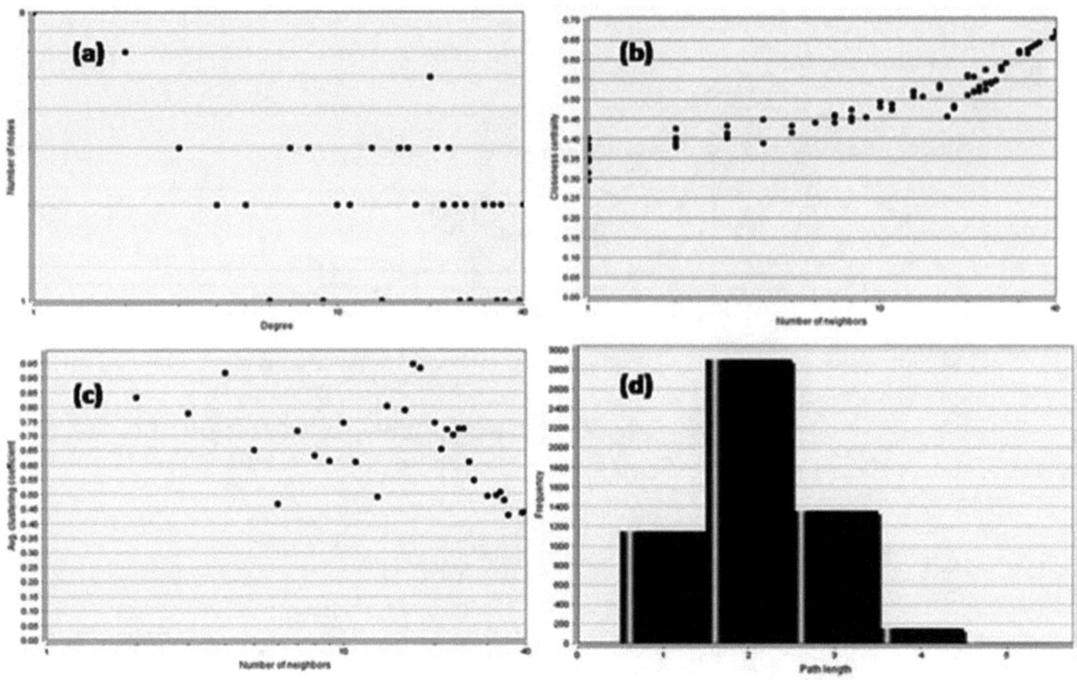

Fig. 7 PPI network analysis of network of inflammatory response and its negative regulation for characteristics of small-world network. (**a**) Distribution of degrees, (**b**) the proximity to the center, (**c**) average aggregation coefficient, and (**d**) distribution of the shortest path [reproduced from ref. 12]

3.7 PPI Network Analysis

1. Analyze the network properties such as distribution of network node degree, distribution of the shortest path, average aggregation coefficient, and proximity to the center [21] for characterizing small-world network [*see* **Note 4**].

In the above stated study, network properties involving distribution of degrees, the proximity to the center, average aggregation coefficient, and distribution of the shortest path were analyzed as shown in Fig. 7a–d.

4 Notes

1. Measures were taken to identify and remove duplicate genes (as there was a possibility of existence of a particular gene in different database).
2. This filtration is required as one miRNA can bind to multiple targets. Therefore, further filtration is required to filter miRNAs based on their sequence alignment with the target and only miRNAs with direct targets are to be included. Hence based on sequence alignment, miRNAs are filtered using the database.

3. Due to strong interaction/connection of gene regulator with transcription factor, STRING database will reveal key gene regulator.
4. Network properties are analyzed for regulating the relationship among various genes.

References

1. Farazi TA, Hoell JI, Morozov P, Tuschl T (2013) MicroRNAs in human cancer. In: MicroRNA cancer regulation. Springer, Dordrecht, p 1–20
2. Liu Z, Sall A, Yang D (2008) MicroRNA: an emerging therapeutic target and intervention tool. Int J Mol Sci 9(6):978–999
3. Drakaki A, Iliopoulos D (2009) MicroRNA gene networks in oncogenesis. Curr Genomics 10(1):35–41
4. Chi SW, Zang JB, Mele A, Darnell RB (2009) Argonaute HITS-CLIP decodes microRNA-mRNA interaction maps. Nature 460:479–486
5. Murakami Y, Tanahashi T, Okada R, Toyoda H, Kumada T, Enomoto M et al (2014) Comparison of hepatocellular carcinoma miRNA expression profiling as evaluated by next generation sequencing and microarray. PLoS One 9:e106314 1-9
6. Kelly AD, Hill KE, Correll M, Hu L, Wang YE, Rubio R et al (2013) Next-generation sequencing and microarray-based interrogation of microRNAs from formalin-fixed, paraffin-embedded tissue: preliminary assessment of cross-platform concordance. Genomics 1:8–14
7. Yoon S, De Micheli G (2006) Computational identification of microRNAs and their targets. Birth Defects Res C Embryo Today 78:118–128
8. Bar M, Wyman SK, Fritz BR, Qi J, Garg KS, Parkin RK et al (2008) MicroRNA discovery and profiling in human embryonic stem cells by deep sequencing of small RNA libraries. Stem Cells 26:2496–2505
9. Li L, Xu J, Yang D, Tan X, Wang H (2010) Computational approaches for microRNA studies: a review. Mamm Genome 21:1–12
10. Akhtar MM, Micolucci L, Islam MS, Olivieri F, Procopio AD (2016) Bioinformatic tools for microRNA dissection. Nucleic Acids Res 44:24–44
11. Christopher AF, Kaur RP, Kaur G, Kaur A, Gupta V, Bansal P (2016) MicroRNA therapeutics: discovering novel targets and developing specific therapy. Perspect Clin Res 7(2):68–74
12. Christopher AF, Gupta M, Bansal P (2016) Micronome revealed miR-19a/b as key regulator of SOCS3 during cancer related inflammation of oral squamous cell carcinoma. Gene 594(1):30–40
13. Szklarczyk D, Franceschini A, Wyder S, Forslund K, Heller D, Cepas JH et al (2015) STRING v10: protein-protein interaction networks, integrated over the tree of life. Nucleic Acids Res 43:D447–D452
14. Hsu SD, Lin FM, Wu WC, Liang C, Huang WC, Chan WL et al (2011) miRTarBase: a database curates experimentally validated microRNA–target interactions. Nucleic Acids Res 39:D163–D169
15. Vergoulis T, Vlachos IS, George PA, Maragkakis GM, Reczko M, Gerangelos S et al (2012) TarBase 6.0: capturing the exponential growth of miRNA targets with experimental support. Nucleic Acids Res 40:D222–D229
16. Xiao F, Zuo Z, Cai G, Kang S, Gao X, Li T et al (2009) miRecords: an integrated resource for microRNA-target interactions. Nucleic Acids Res 37:D105–D110
17. Jiang Q, Wang Y, Hao Y, Juan L, Teng M, Zhang X et al (2009) miR2Disease: a manually curated database for microRNA deregulation in human disease. Nucleic Acids Res 37:D98–D104
18. Shannon P, Markiel A, Ozier O, Baliga NS, Wang JT, Ranage D et al (2003) Cytoscape: a software environment for integrated models of biomolecular interaction networks. Genome Res 13:2498–2504
19. Agarwal V, Bell GW, Nam J, Bartel DP (2015) Predicting effective microRNA target sites in mammalian mRNAs. E Life 4:e05005 1-38
20. Shirdel EA, Xie W, Mak TW, Jurisica I (2011) Navigating the micronome–using multiple microRNA prediction databases to identify signalling pathway-associated microRNAs. PLoS One 6:e17429 1-17
21. Assenov Y, Ramírez F, Schelhorn SE, Lengauer T, Albrecht M (2008) Computing topological parameters of biological networks. Bioinformatics 24:282–284

Chapter 15

Method for Detection of miRNAs in Non-Model Organisms with Unreported Database

Parveen Bansal, Ashish Kumar, Sudhir Chandna, Malika Arora, and Renu Bansal

Abstract

Non-model organisms are studied very frequently, as a simple accessible and convenient system to investigate the role of miRNAs in particular aspect of biology or disease. However, the unavailability of the annotated genome and hence miRNA database of these non-model organisms pose a major constraint for using them more efficiently. Here, we describe a new method to identify miRNAs in non-model organisms without complex sequencing strategies and using miRNAs from close relative organisms as proxy/reference sequences.

Key words miRNA, Non-model organisms

1 Introduction

Micro-RNAs comprise 18–22 bp RNA family, which are highly conserved among evolutionarily related species, and many even have homologies in distantly related species and represent a unique mechanism for controlling gene expression via binding to mRNA and either block transcription or promote mRNA degradation. Like mRNAs, micro-RNAs are also transcribed by RNA pol II in the nucleus. The biogenesis of miRNAs is compartmentalized into two locations, nuclear and cytoplasmic. After initial transcription as primary micro-RNAs ranging from few hundreds to thousands nucleotides [1, 2], these pri-miRNAs are first processed in nucleus by microprocessor complex, comprised of the RNase III enzyme, Drosha and the double-stranded RNA binding domain (dsRBD) protein DGCR8 and generate 60–70 nt precursor miRNA (pre-miRNA) intermediates [3–5]. The resulting pre-miRNA have 5′ phosphate and 3′ hydroxy termini, and two- or three-nucleotide 3′ single-stranded overhanging ends, all of which are characteristics of RNase III cleavage of dsRNA. The pre-miRNA is then exported

Ulf Andersson Vang Ørom (ed.), *miRNA Biogenesis: Methods and Protocols*, Methods in Molecular Biology, vol. 1823,
https://doi.org/10.1007/978-1-4939-8624-8_15,

from nucleus to cytoplasm by Exportin 5/RanGTP, which specifically recognizes the characteristic end structure of pre-miRNAs [6–9]. In the cytoplasm, a second RNase III, Dicer, together with its dsRBD protein partner, process this pre-miRNA into 21 nucleotide long RNA duplex [10–14]. In the RNA duplex, the strand having less thermodynamically stable 5′ end will emerge into mature micro-RNA [15]. The mature miRNA then assembled into the effectors complex with the help of Argonaute (Ago) together with other probable partners like Gemin3, Gemin4, vasa intronic gene (VIG), Fragile X mental retardation protein (FMRP), and a staphylococcal nuclease and Tudor-SN. The RISC-loaded mature miRNAs are protected from degradation by Argonaute proteins [16, 17] and can bind to the 3' UTRs of their respective mRNA targets.

The miRNA expression pattern affects many cellular pathways; it depends on the microenvironment or cell/tissue type, and may vary with the extent of stress as well as time. To explore the role of miRNAs in the regulation of any cellular pathway miRNA mimics and inhibitors are commonly used to alter their expression and study the effect on modulated genes/pathways. However, sometimes it is difficult to study the role of miRNAs in vivo in many organisms due to complexity of the organism's genome or difficulties in maintaining, breeding or relevant ethical issues. Some non-model organisms show close evolutionary relationship with model organisms including human system. Hence, non-model organisms sometimes represent the best option to overcome such limiting issues. While study on the non-model organisms may be required to establish the involvement of miRNAs in many diseases and/or biomarkers and reveal other cellular/molecular pathways, often these studies are limited by unavailability of the requisite miRNA database or even genome database. Usually researchers use miRNA sequences from closely related species as proxy references to facilitate miRNA expression studies in non-model species if their genomes are not available. Commonly, sequencing the genome of non-model organisms and mapping their genome with the model organisms is a popular approach to identify relevant miRNA sequences [18, 19]. However, the level of error this is likely to introduce could be difficult to ascertain, as this may get influenced by evolutionary distance between the proxy reference and the species of interest. In many studies, detection of known/conserved miRNAs and their expression levels is a priority rather than discovering novel miRNAs. At the same time it is also important, even if miRNA sequences from closely related species are used as reference, to establish if that sequence is present in the species in question and if so, that miRNA shares the exact sequence similarity [20].

Identification of conserved miRNAs using method described here is based on RNase protection assay. Using this method it is possible to find out if a particular miRNA is present in non-model organism and if so whether it shares the exact similar sequence with the reference organism. Using biotin-labeled single-stranded RNA probes for RNase protection assay and chemiluminescence detection of protected miRNA, it is possible to identify and establish a particular miRNA sequence in the non-model organisms.

2 Materials

All the micro tips and tubes used in this method should be treated with 0.1% diethyl pyrocarbonate (DEPC) for at least 2 h and further autoclaved for 15 min to remove any traces of DEPC. The work bench and pipettes should also be wiped with RNase Zap to avoid any RNase contamination.

2.1 Design of DNA Oligonucleotide Template for Probe Construction

1. For in vitro construction of probe for miRNA identification, a DNA sequence is required with some additional conserved bases (*see* **Note 1**).
2. The DNA sequence should have the same sequence as that of miRNA (from a reference organism) except the "U"s should be replaced with "T"s. Additionally, 8 nucleotide sequence (5′-CCTGTCTC-3′) should be added to the 3′ end of the single-stranded DNA template, which is complementary to the T7 Promoter Primer and will allow hybridization of DNA oligonucleotide template with the T7 Promoter Primer.
3. As an example, template design for hsa-miR-34a sequence should be: >hsa-miR-34a-5p: 5′-UGGCAGUGUCUUAGCUGGUUGU-3′ DNA template: 5′-TGGCAGTGTCTTAGCTGGTTGT**CCTGTCTC**-3′.
4. An additional single-stranded DNA sequence (T7 promoter primer) is required that must have T7 promoter sequence at its 5′ end followed by any 30–40 bp scrambled sequence.

2.2 Buffers and Reagents

1. Urea PAGE gel: 7.2 g 8 m Urea, 1.5 mL 10× TBE, 5.6 mL 40% Acrylamide (19:1 ratio of Acrylamide: bisacrylamide). Make the volume up to 15 mL using water. Mix well, add 75 μL of ammonium persulfate and 15 μL TEMED just before pouring in the casting plates (glass plates) with 1 mm spacer (*see* **Note 2**).
2. 20 μM stock of each of Adenosine triphosphate (ATP), Guanosine triphosphate (GTP), Cytosine triphosphate (CTP), and Biotin-labeled uridine triphosphate (Biotin-16 UTP).
3. Hybridization buffer (2×): 10 mM Tris-HCl (pH 7.5), 1 mm EDTA, 2 M NaCl.

4. Transcription Buffer: 1× buffer: 50 mM Tris-HCl, pH 7.5, 15 mM $MgCl_2$, 5 mM dithiothreitol (DTT), 2 mM spermidine. Make 10× stock and store at −20 °C.
5. 10× TBE: 0.9 M Tris base, 0.9 M Boric acid, 20 mM EDTA (*see* **Note 3**).
6. Washing buffer: 20 mM Tris-HCl, 500 mM NaCl, (pH 7.5) (TBS buffer); 0.1% Tween-20.
7. Blocking Buffer: 2% BSA in washing buffer.

3 Methods

All the procedures should be carried out at room temperature unless otherwise specified.

3.1 Mature miRNA Isolation

1. For mature miRNA isolation, any preferred method/kit can be used that can enrich mature miRNA.
2. Alternatively, total RNA isolated using standard Trizol method can be used for the mature miRNA extraction/enrichment by Urea PAGE.
3. Prepare 15% UREA PAGE gel and pre-run the gel for 30 min in 0.5× TBE buffer at 100 V.
4. Mix 10–20 μg of total RNA with loading dye (contains xylene cyanol) and heat at 95 °C for 5 min and cool on ice for 5 min.
5. Rinse the wells thoroughly with 0.5× TBE and load the total RNA.
6. Load the samples at the bottom of the gel pocket and run until xylene cyanol dye front reaches ¾ of the gel length.
7. Open the gel cast/plates and stain the gel with 0.5 μg/mL ethidium bromide (or any other stain) and cut the gel piece from the mid of the xylene cyanol to the half of the bromophenol blue dye. Generally, xylene cyanol run at about 30 bp on denaturing Urea PAGE.
8. Crush the gel piece in 500–1000 μL of TE (or elution buffer of choice) and leave it overnight at 4 °C.
9. Next day centrifuge briefly to remove the gel pieces and take supernatant to fresh tube.
10. Add 95% ethanol to the final concentration of 80% and 0.1 volume of 3 M NaOAc (pH 5.5) to the supernatant and incubate at −20 °C for 2-4 h.
11. Centrifuge at 12,000 RPM (23182 × *g*) for 15 min at 4 °C.
12. Wash the pellet with 70% ethanol and resuspend 50 μL of TE buffer.

3.2 In Vitro Probe Construction

For in vitro transcription of probes, the target DNA template (similar as mature miRNA sequence with U replaced with T and additional 8 nucleotides at 3′ end) should be first converted to double strand DNA (dsDNA) with T7 promoter at 5′ end.

3.2.1 dsDNA Template Synthesis for In Vitro Transcription

1. Dilute template DNA into 100 μM stock in nuclease-free water.
2. Dilute scrambled DNA sequence that has T7 promoter sequence at its 5′ end, to 100 μM stock.
3. Prepare the following reaction: using 2 μL of oligonucleotide template and 2 μL of scrambled sequence (having T7 promoter sequence) along with 6 μL of hybridization buffer.
4. Heat the mixture to 70 °C for 5 min, then leave at room temp for 5 min.
5. Mix 20 μM of each ATP, GTP, CTP, and UTP (no biotin) to final volume of 20 μM. That will be 10× dNTP stock.
6. Add following to the hybridization mix from **step 4**: 2 μL of dNTP mix (10×), 2 μL of Exo-Klenow enzyme, 2 μL of exo-Klenow buffer, and 4 μL of nuclease-free water, mix gently, and incubate at 37 °C for 30–45 min to generate dsDNA template.
7. dsDNA template can be stored at −20 °C.

3.2.2 In Vitro Transcription

1. Mix 1 μL of dsDNA template (from above step), 2 μL of 10× Transcription buffer, 1 μL each of ATP, GTP, CTP and Biotin UTP, 2 μL of T7 RNA polymerase. Make the final volume up to 20 μL by adding 11 μL of RNase-free water, at room temperature.
2. Gently mix the reaction by pipetting up and down several times.
3. Incubate the reaction at 37 °C for 30 min.
4. Add 1 μL of DNase I and gently pipette up and down to remove the template DNA.
5. Incubate again at 37 °C for 10 min.

3.3 Gel Purification of Transcribed RNA Probes

1. Prepare 15% denaturing PAGE gel with 1 mm spacer (*see* **Note 4**).
2. Pre-run for 30 min at 80–100 V and wash the wells thoroughly with 0.5% TBE before loading the sample.
3. Mix the probes with gel loading dye (with bromophenol blue) and heat at 95 °C for 5 min.
4. Load the samples carefully from the bottom of the gel pocket on a denaturing 15% polyacrylamide gel. Avoid introducing air bubbles. Electrophorese at 20–25 mAmp.

5. Stop electrophoresis when bromophenol blue dye front reaches at the bottom of the gel. Bromophenol blue dye migrates at about 10 bp on 15% denaturing polyacrylamide gel.
6. Stain the gel with 0.5 μg/mL ethidium bromide in 1× TBE for 5 min with intermittent gentle shaking.
7. Wash the gel for 2 min with 1× TBE.
8. Visualize using a UV transilluminator.
9. The pattern followed by dsDNA template and probes on 15% denatured polyacrylamide gel is shown in Fig. 1.

3.4 RNase Protection Assay

For the detection of miRNAs in non-model organisms, it is also required to use positive and negative controls. For positive controls, mature miRNAs can either be extracted from that reference organism or miRNA mimic (with the same sequence) can be used. For negative control, no miRNA should be used with ssRNA biotinylated probes. Additionally, one "no RNase digestion" should also be used with **NO** miRNA sample and **NO** RNase digestion, though total number of samples will be four in this case.

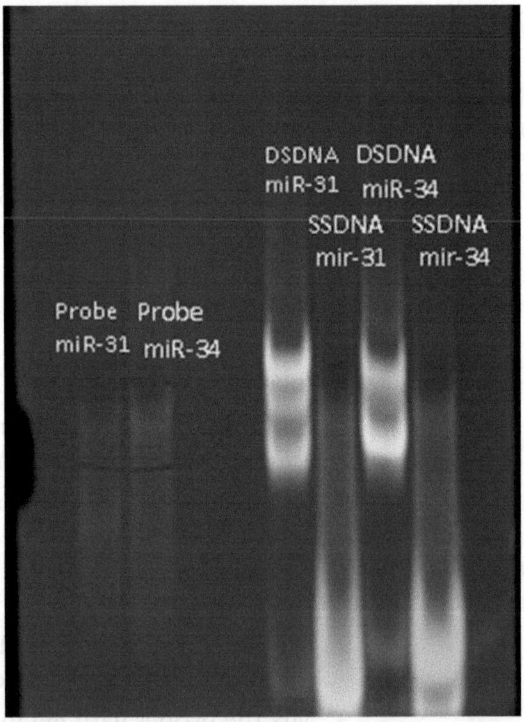

Fig. 1 Gel image representation of single-stranded miRNA template (ssmiR-31 and ssmiR-34 sequences in DNA format), dsDNA, and probes from two different miRNAs, miR-31 and mir-34. *Bombyx mori* miRNAs were used as reference sequences for dsDNA template synthesis and in vitro probe construction for the identification of these miRNAs in non-model organism *Spodoptera frugiperda*

3.4.1 Hybridization of Mature miRNAs with ssRNA Probe

1. Mix the following components in a PCR tube: 10 μL of hybridization buffer, 2 μL of in vitro transcribed probe, 1–2 μg of mature miRNA (in 8 μL) and make the final volume up to 20 μL with nuclease-free water (*see* **Note 5**).
2. For no miRNA control, we recommend to use yeast RNA in equal concentration as that of miRNA reaction (i.e., 1–2 μg).
3. Mix the reaction by pipetting up and down several times and centrifuge briefly to collect the reaction mix at the bottom of the tube.
4. Heat the reaction mix at 95 °C for 5 min to break any secondary structure.
5. Incubate the reaction at 37 °C for 12–16 h (or overnight) in thermo cycler. It is also recommended to optimize incubation time for each set of miRNA and probe.

3.4.2 RNase Digestion of Hybridized Probe and miRNA

For RNase digestion, a mix of RNase A/T1 is recommended. The RNase A specifically hydrolyzes RNA at C and U residues, while RNase T1 specifically hydrolyzes RNA at G residues.

1. Dilute the RNase A/T1 mix in RNase digestion buffer in 1:100 dilution. However the concentration of RNase A/T1 depends upon the amount of total mature miRNAs used for hybridization.
2. Remove the hybridization reaction from the thermocycler (From above step) and centrifuge briefly to collect the entire sample at the bottom of the tube.
3. Add 100 μL of diluted RNase A/T1 mix to each tube except "no RNase digestion" negative control. Add 100 μL of RNase A/T1 digestion buffer in that sample.
4. Mix the reaction by pipetting up and down several times.
5. Incubate the reaction for 45 min at 37 °C in a thermocycler. During incubation all the mature miRNAs that are not protected by probes should be degraded by RNase A/T1.
6. After incubation, transfer the reaction to 1.5 mL tube and add 450 μL of 95–100% ethanol and 60 μL of 3 M NaOAc.
7. Mix thoroughly by pipetting up and down several times.
8. Incubate the reaction at −20 °C for 2–3 h.
9. Centrifuge at 12,000 rpm (23182 × *g*) for 15 min at 4 °C.
10. Wash the pellet (invisible) with 70% ethanol.
11. Dry the tube for 5 min in a laminar flow.
12. Dissolve the pellet in 10 μL of gel loading dye containing bromophenol blue and xylene cyanol.

3.5 Separation of Protected miRNA

For separation of hybridized probe and miRNA, 15% denatured polyacrylamide gel should be used.

1. Heat all the samples (including one positive control and two negative controls), dissolved in gel loading dye for 95 °C for 5 min before loading. "No RNase digestion" sample should be diluted 1:10 times, as it holds the undigested probe and might give very high signals during chemiluminescent blot development.
2. Pre-run the gel for 30 min in 0.5% TBE.
3. Rinse the wells thoroughly with 0.5% TBE before loading the samples.
4. Load the samples at the bottom of the gel pocket using 10 μL pipette.
5. Run the gel at 100 V till the bromophenol blue dye front reaches at the bottom of the gel.

3.6 Chemiluminescent Detection and Identification of Protected miRNA

After the gel electrophoresis run is completed, dismantle the electrophoresis assembly and remove the gel from the glass plates. Soak the nylon membrane in 0.5% TBE buffer for 5 min while the gel is running. TBE buffer (0.5%) should also be chilled for at least 2 h at 4 °C.

1. Place the gel on a clean transfer sponge in electrophoretic transfer unit.
2. Place wet nylon membrane on the gel so that it covers all the lanes loaded with the samples.
3. Sandwich the gel and nylon membrane with another clean transfer sponge.
4. Place this assembly in the standard tank transfer apparatus for mini gels (8 × 8 × 0.1 cm) and fill the tank to the top with chilled 0.5% TBE.
5. Transfer at 400 mA (~35 V) for 45 min at 4 °C.
6. After the transfer is complete, place the membrane with the bromophenol blue dye side up on a dry paper towel to absorb buffer on the membrane surface but at the same time do not let the membrane dry (*see* **Note 6**).

   Cross-link the transferred (protected) miRNA and probe using a UV hand lamp. Place the UV hand lamp, equipped with 254 nm bulb, at a distance of about 5 mm from the membrane for about 5 min (*see* **Note 7**).

7. After the membrane is cross-linked, place the membrane in a trough, slightly bigger than the size of the membrane.
8. Add sufficient amount of blocking buffer to submerge the membrane and incubate for 45 min at room temperature with gentle shaking.

9. Discard the blocking buffer and add freshly diluted Streptavidin HRP conjugate (approximately 0.5 μg/mL) in blocking buffer (*see* **Note 8**).
10. Incubate the membrane for 1 h at room temperature on a shaking platform.
11. Discard the Streptavidin HRP conjugate solution and wash the membrane three times with washing buffer for total 30 min with gentle shaking.
12. Place the membrane in a new trough and add substrate to produce the signals which can be captured on X-ray film or using any charge-coupled device (CCD) camera-based digital imaging instruments.

Figure 2 represents the chemiluminescent blotting and identification of miRNA-31 in non-model organism *Spodoptera frugiperda*. miR-31 showed exact sequence similarity with that of reference organism (*Bombyx mori*).

4 Notes

1. As shown in Fig. 2, miR-31 was identified in *Spodoptera frugiperda* and showed exact sequence similarity with *Bombyx mori*, as the complete miRNA-31 sequence was protected. However, some miRNAs may have multiple members of the same family (i.e., miR-34a/b etc.) in one organism/species but may not

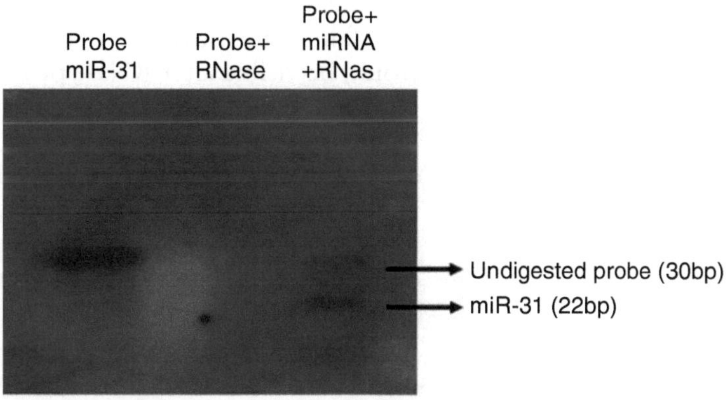

Fig. 2 Representation of miR-31 chemiluminescent blotting from non-model organism *Spodoptera frugiperda. Bombyx mori* miR-31 sequence was used as reference. Left lane (extreme left) represents the undigested probe ("no RNase digestion Control") and signifies a 30 bp band (22 bp miRNA sequence and 8 bp T7 promoter sequence). Middle lane shows the complete digestion of the probe after RNase digestion. Right lane represents the protected miR-31 (≈22 bp) sequence from *Spodoptera frugiperda* with some undigested probe [reproduced from ref. 20]

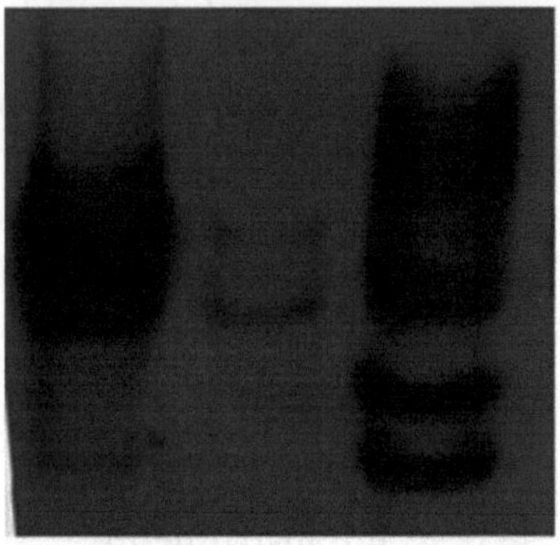

Fig. 3 Chemiluminescent blotting of miR-34 in *Spodoptera frugiperda* (Sf9 cells) samples using *Bombyx mori* reference sequence. Left lane represents the undigested probe. Middle lane shows RNase A/T1 digestion. Right lane shows two bands along with undigested probe. The upper band represents the ~22 bp sequence, and hence confirms the presence and sequence similarity of miR-34 sequence in *Spodoptera frugiperda.* The additional lower (partially protected) band indicates the presence of other miR-34 family member in *Spodoptera frugiperda*, with partial sequence similarity

present in the closely related organism. Thus, if the miRNAs have multiple members in the reference organism, all the members should be used as a template for the identification of miRNA members in the non-model organism. However, if in the reference organism, only one form of a particular miRNA is present and the non-model organism represents more than one family members of that miRNA, more than one band would be visible after RNase protection and chemiluminescent blotting. Figure 3 is a very good representation of this case, where miR-34 sequence is used from *Bombyx mori*, which possesses the only form of miR-34. Hybridizing the probes (generated from reference sequence) with miRNA extracted from *Spodoptera frugiperda* represented two protected bands and one of these bands moving faster than 22bp represents that apart from full length miRNA (upper band ≈22 bp), another miRNA sequence was also partially protected (lower band) and may be a member of the same miRNA family. That partially protected band may result from one base difference between the reference miR-34 sequence and that from *Spodoptera frugiperda.*

2. It is very important to be cautious while using acrylamide and ethidium bromide. Always use gloves and mask while working with these chemicals to avoid any exposure.
3. It is good idea to use TBE buffer instead of TAE. TBE buffered gels yield sharper bands in comparison to TAE, particu-

larly when using small-sized RNA/DNA/miRNAs, and can be run at higher voltages.

4. The sharpness of the bands also depends on the thickness of the gel. Thinner gels such as 0.75 mm show better results than 1.5 mm thick gels. Also shape of bands can be distorted, if the sample is not applied equally into the gel pocket. Including additional 10% glycerol in the gel loading dye helps during the gel loading, and the sample sinks into the well more easily.
5. The amount of mature miRNA used in the hybridization reaction can fluctuate depending upon the abundance of the miRNA to be identified. Total amount of mature miRNA from 500 ng to 10 μg have been tested successfully by our lab.
6. There should be no dye remaining in the gel after transfer is complete. Also it is always a good idea to mark the position and orientation of the membrane as after removing the membrane from the gel the down side (touching the gel) should be placed upwards. Cutting the upper right corner of the membrane is always a good idea.
7. Excessive exposure to UV will cause degradation of the RNA.
8. The dilutions of streptavidin HRP conjugate should be optimized.

References

1. Cai X, Hagedorn CH, Cullen BR (2004) Human microRNAs are processed from capped, polyadenylated transcripts that can also function as mRNAs. RNA 10(12):1957–1966
2. Lee Y, Kim M, Han J, Yeom KH, Lee S, Baek SH, Kim VN (2004) MicroRNA genes are transcribed by RNA polymerase II. EMBO J 23(20):4051–4060
3. Gregory RI, Yan KP, Amuthan G, Chendrimada T, Doratotaj B, Cooch N, Shiekhattar R (2004) The microprocessor complex mediates the genesis of microRNAs. Nature 432(7014): 235–240
4. Denli AM, Tops BB, Plasterk RH, Ketting RF, Hannon GJ (2004) Processing of primary microRNAs by the microprocessor complex. Nature 432(7014):231–235
5. Han J, Lee Y, Yeom KH, Kim YK, Jin H, Kim VN (2004) The Drosha-DGCR8 complex in primary microRNA processing. Genes Dev 18(24):3016–3027
6. Yi R, Qin Y, Macara IG, Cullen BR (2003) Exportin-5 mediates the nuclear export of pre-microRNAs and short hairpin RNAs. Genes Dev 17(24):3011–3016
7. Bohnsack MT, Czaplinski K, GÖRLICH D (2004) Exportin 5 is a RanGTP-dependent dsRNA-binding protein that mediates nuclear export of pre-miRNAs. RNA 10(2):185–191
8. Lund E, Güttinger S, Calado A, Dahlberg JE, Kutay U (2004) Nuclear export of microRNA precursors. Science 303(5654):95–98
9. Zeng Y, Cullen BR (2004) Structural requirements for pre-microRNA binding and nuclear export by Exportin 5. Nucleic acids Res 32(16):4776–4785
10. Bernstein E, Caudy AA, Hammond SM, Hannon GJ (2001) Role for a bidentate ribonuclease in the initiation step of RNA interference. Nature 409(6818):363–366
11. Grishok A, Pasquinelli AE, Conte D, Li N, Parrish S, Ha I et al (2001) Genes and mechanisms related to RNA interference regulate expression of the small temporal RNAs that control C. elegans developmental timing. Cell 106(1):23–34
12. Hutvágner G, McLachlan J, Pasquinelli AE, Bálint É, Tuschl T, Zamore PD (2001) A cellular function for the RNA-interference enzyme dicer in the maturation of the let-7 small temporal RNA. Science 293(5531):834–838
13. Chendrimada TP, Gregory RI, Kumaraswamy E, Norman J, Cooch N, Nishikura K, Shiekhattar R (2005) TRBP recruits the dicer

complex to Ago2 for microRNA processing and gene silencing. Nature 436(7051): 740–744

14. Saito K, Ishizuka A, Siomi H, Siomi MC (2005) Processing of pre-microRNAs by the Dicer-1–loquacious complex in drosophila cells. PLoS Biol 3(7):e235
15. Tomari Y, Zamore PD (2005) Perspective: machines for RNAi. Genes Dev 19(5):517–529
16. Meister G, Landthaler M, Patkaniowska A, Dorsett Y, Teng G, Tuschl T (2004) Human Argonaute2 mediates RNA cleavage targeted by miRNAs and siRNAs. Mol Cell 15(2):185–197
17. Rand TA, Petersen S, Du F, Wang X (2005) Argonaute2 cleaves the anti-guide strand of siRNA during RISC activation. Cell 23(4):621–629
18. Kayvan E, Mazhar H, Sassan A (2013) Identification of microRNAs from Plutella xylostella larvae associated with parasitization by Diadegma semiclausum. Insect Biochem Mol Biol 43(4):309–318
19. Wu WQ, Ren Q, Li C, Wang Y, Sang M, Yi Z et al (2013) Characterization and comparative profiling of MicroRNAs in a sexual dimorphism insect, Eupolyphaga sinensis Walker. PLoS One 8(4):e59016 1–8
20. Kumar A, Ghosh S, Chandana S (2015) Evidence for microRNA-31 dependent Bim-Bax interaction preceding mitochondrial Bax translocation during radiation-induced apoptosis. Sci Rep 5:1–14

Chapter 16

Detection and Verification of Mammalian Mirtrons by Northern Blotting

Mohammad Farid Zia and Alex S. Flynt

Abstract

microRNAs (miRNAs) have vital roles in regulating gene expression—contributing to major diseases like cancer and heart disease. Over the last decade, thousands of miRNAs have been discovered through high throughput sequencing-based annotation. Different classes have been described, as well as a great dynamic range of expression levels. While sequencing approaches provide insight into biogenesis and allow confident identification, there is a need for additional methods for validation and characterization. Northern blotting was one of the first techniques used for studying miRNAs, and remains one of the most valuable as it avoids enzymatic manipulation of miRNA transcripts. Blotting can also provide insight into biogenesis by revealing RNA processing intermediates. Compared to sequencing, however, northern blotting is a relatively insensitive technology. This creates a challenge for detecting low expressed miRNAs, particularly those produced by inefficient, non-canonical pathways. In this chapter, we describe a strategy to study such miRNAs by northern blotting that involves ectopic expression of both miRNAs and miRNA-binding Argonaute (Ago) proteins. Through use of epitope tags, this strategy also provides a convenient method for verification of small RNA competency to be loaded into regulatory complexes.

Key words miRNA, Mirtron, Tailed-mirtron, Low-abundance detection, RNA immunoprecipitation, Northern blot

1 Introduction

Small non-coding RNAs are common features of eukaryotic transcriptomes that require specialized detection methods due to their small size. Moreover many species are present in low abundance, which further complicates detection [1]. microRNAs are an ~23 nucleotide (nt) class of small RNAs that have vital roles in gene regulation and their dysregulation is associated with numerous diseases [2, 3]. Animal microRNAs are a highly diverse class of genes. Some miRNAs are ancient, predating the birth of metazoans, while others are newly evolved and species specific [4]. Accordingly, microRNAs are a highly diversified class of transcript [5]. High throughput sequencing has been instrumental in small RNA discovery [6]. Indeed, thousands of miRNAs have been discovered

Ulf Andersson Vang Ørom (ed.), *miRNA Biogenesis: Methods and Protocols*, Methods in Molecular Biology, vol. 1823,
https://doi.org/10.1007/978-1-4939-8624-8_16,

and deposited into databases [7]. Yet, compared to the number of miRNAs annotated, there have been very few functional studies, suggesting there is a need for additional strategies for detection.

miRNA biogenesis typically involves cleavage from hairpin RNAs precursors (pre-miRNAs). For most miRNAs, the Drosha/DGCR8-containing microprocessor complex excises pre-miRNA from primary transcripts [8]. Final processing is then carried out by Dicer, yielding miRNA/miRNA* duplexes [9]. One strand then is loaded into Argonuate (Ago) proteins where it participates in targeting and repression of transcripts [10]. Beyond this canonical pathway, there are many atypical pathways. Multiple varieties of miRNAs have been found to bypass Drosha/Dgcr8 processing like mirtrons, shRNAs, snoRNA-derived miRNAs, tRNA-derived miRNAs, and rRNA-derived miRNAs (Fig. 1) [11–15]. In this chapter, we will focus on the detection of mirtrons. They are derived from small introns that after splicing and debranching can fold into small hairpins [2, 16, 17].

Hundreds of mirtrons have been discovered in the human genome through inspecting hairpins coincident with splice sites from which small RNA sequencing reads are derived [18]. Mirtrons are classified in three categories: conventional, 3′-tailed, and 5′-tailed. In conventional, both hairpin ends correspond to splice sites and after splicing can become dicer substrates. In "tailed" varieties only one side of the hairpin is at a splice site, while the other side has a stretch of nt's, or tail, between the hairpin base and splice site [19, 20]. Studies in *Drosophila* have also found that mirtrons are commonly modified by Tailor, a TUTase (terminal Uridyltransferase), which uridylates hairpins to inhibit their biogenesis by abrogating Dicer processing [21, 22]. Thus, mirtrons more so than other types of miRNAs are less robustly expressed than canonical miRNAs.

Comparing expression of human canonical miRNAs (deep conserved or newly evolved) to mirtrons inferred from small RNA sequencing data meta-analysis reveals that mirtrons have on average several fold lower abundance (Fig. 2) [23]. Furthermore, the distribution of mirtrons is bimodal, showing that many species accumulate at the lowest levels of expression observed for miRNAs (Fig. 2). The type of mirtron in the lowest abundance population are tailed varieties, consistent with the requirement for additional processing during tail removal. Therefore, detection strategies capable of detecting tailed-mirtrons via northern blotting will be applicable for studying all types of low abundance miRNAs.

Northern blotting was the first technique used to detect a miRNA transcript [24]. Many techniques have been developed in the interim which frequently rely on reverse transcription and PCR, which can introduce bias, artifacts, and mostly can't distinguish between biogenesis intermediates [25, 26]. The challenge in using northern blotting is relatively low sensitivity, making

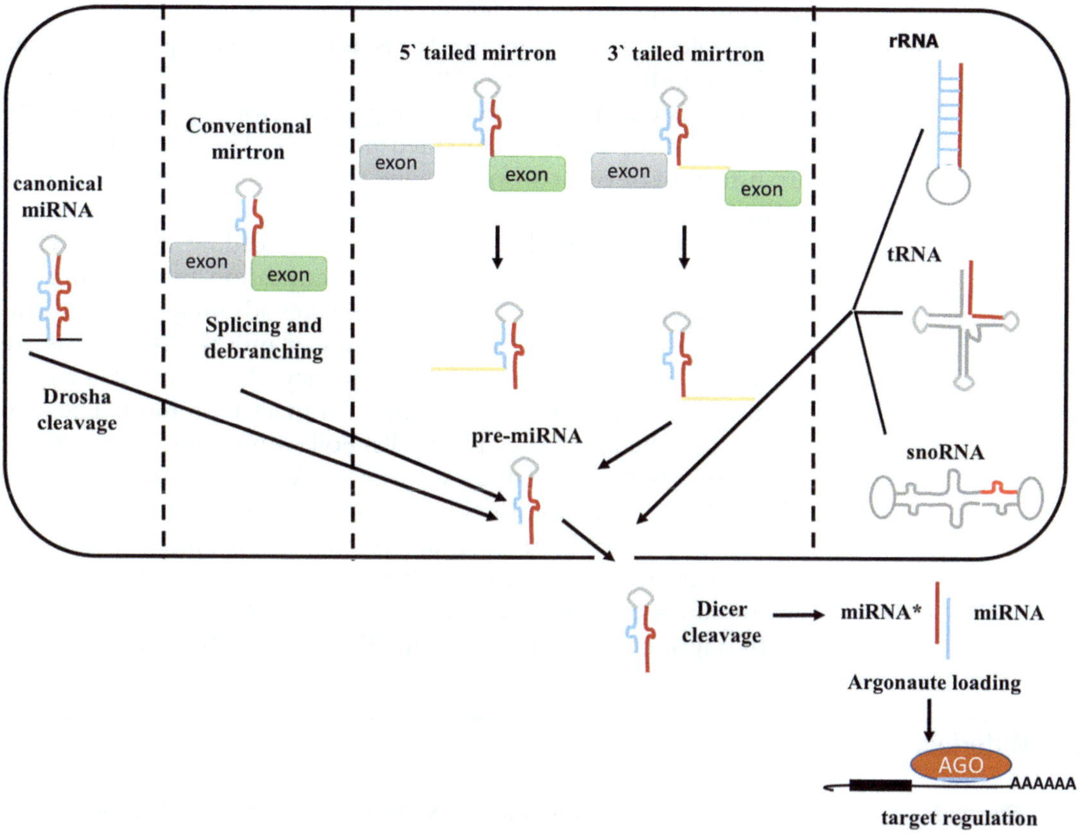

Fig. 1 Classes of miRNAs defined by biogenesis pathways

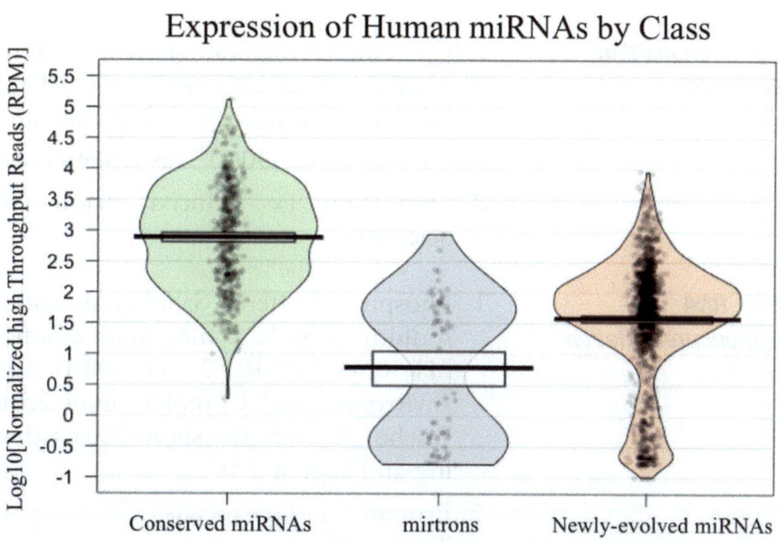

Fig. 2 RDI plot showing the distribution of normalized (Reads per million) miRNA expression separated by class. Values were taken from miRbase

detection of tailed-mirtrons problematic. Despite this difficulty, it is possible to detect tailed-mirtrons using ectopic expression of the miRNA and Ago proteins [18].

In this chapter, we describe a protocol for applying this technique to study mammalian tailed-mirtrons. This process involves transfection of two constructs, one to express the mirtron and the other to express a myc-tagged human Ago2 (Fig. 3a). In this system, the mirtron is cloned into the open-reading frame of GFP, producing a fluorescent reporter for verification that splicing has occurred efficiently. This strategy also takes advantage of a cell culture phenomena where increased rates of miRNA biogenesis are observed in cells grown to high confluency [27]. Thus, it is critical to use a cell line that does not exhibit contact-inhibition. The protocol described here uses Hela cells. Following transfection, RNAs can either be bulk isolated by a routine RNA purification method like TRIzol extraction or through immunoprecipitation targeting the myc epitope fused to Ago2. Northern blotting, as previously described, involving urea-PAGE separation and transfer to nylon membranes can then be used to visualize RNAs in conjunction with oligonucleotide probes (Fig. 3b) [28] (*see* **Note 1**).

2 Materials

2.1 Plasmid Constructs

Mirtrons are amplified and inserted into expression vectors within opening reading frame of GFP [29]. The protocol also utilizes a myc-tagged hAgo2 (Addgene: Plasmid #19872). Plasmid DNAs are isolated via commonly used commercial purification kits like the Qiagen®Endo-free Maxi Prep Kit.

2.2 Transfection Reagents

1. Cell culture medium: DMEM (Dulbecco's Modified Eagle's Medium) supplemented with 10% FBS and 1% penicillin/Streptomycin using standard cell culture methodology [30].
2. Lipofectamine 3000® Invitrogen reagent. Keep at 4 °C [30].
3. Opti-Mem 1× reduced serum medium (Thermo Fisher Scientific).

2.3 RNA Immunoprecipitation

1. Phosphate buffered Saline, lysis buffer (5% Glycerol, 150 mM sodium chloride, 2 mM Magnesium Acetate, 20 mM Tris 7.5 pH, 0.5% NP-40, 2 mM DTT, 40 units/mL RNase Out® Invitrogen, and Complete mini Protease Inhibitor Cocktail® Roche). Lysis buffer should be used within several days of mixing and kept at 4 °C.
2. Protein G Dynabeads® Invitrogen, magnetic stand, and Tween-20.
3. Anti-myc (clone 9E10, Thermofisher) and a control IgG (Jackson Immunoresearch) antibodies.

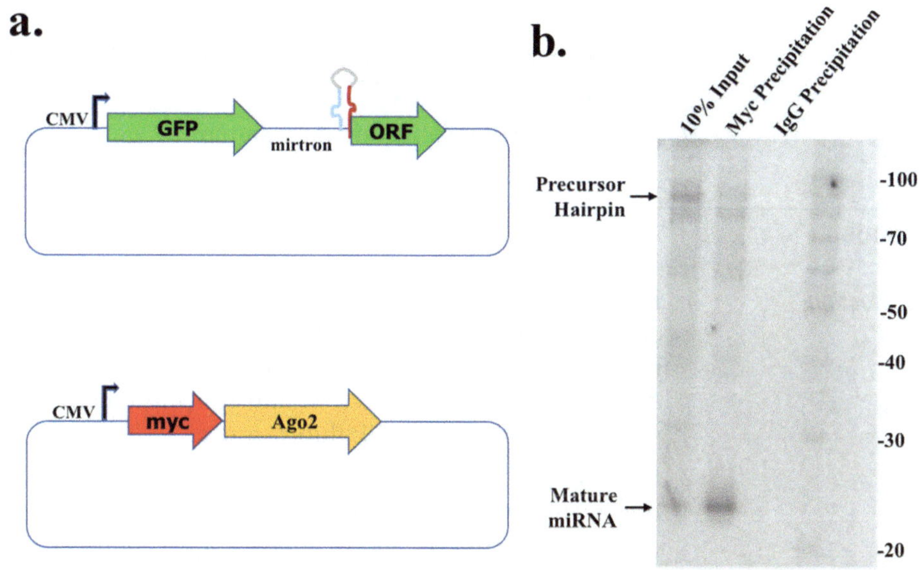

Fig. 3 System for detecting and verifying mammalian mirtrons. (**a**) Constructs for co-transfection. Top vector configured to couple mirtron splicing to GFP expression, by interrupting the GFP ORF with the mirtron-containing intron. Bottom vector for ectopic expression of Ago2-myc. (**b**) Northern Blotting reveals expression of miR-5010, a 5′-tailed mirtron, using the vectors in (**a**) in conjunction with an immunoprecipitation procedure

4. Bead preparation and RNA isolation: TRIzol® Invitrogen, or 400 mM sodium chloride, acid phenol chloroform (Ambion), glycogen co-precipitant, isopropanol, 70% ethanol, and sodium acetate pH 5.3.

2.4 Northern Blotting

1. RNA loading buffer II, Tris-Boric Acid EDTA (TBE) running buffer [31].
2. 12% urea PAGE gel, Ammonium persulfate (10% solution in water) (APS), and N,N,N,N′ Tetramethylethylenediamine (TEMED) [32].
3. Positively Charged Nylon membrane (Hybond-N, GE-lifesciences), Filter paper, Semi-dry electroblotting apparatus [33].
4. DNA probe hybridization buffer: 5× SSC, 1 mM EDTA, 2× Denhardt's, 1% SDS, 2% Dextran Sulfate, 30 μg/mL. Combine all components and boil until completely dissolved.
5. Probes: T4 polynucleotide kinase (PNK) (NEB), DNA oligonucleotides complementary to the miRNA of interest, gamma-P^{32}ATP 6000 μCi/mL (Perkin Elmer), and Sephadex G-25 columns (GE-Life Sciences).
6. Decade Marker (Ambion).
7. Wash Buffers: High stringency (2× SSC, 0.1% SDS) and low stringency (0.2× SSC, 0.1% SDS).

8. LNA probe hybridization buffer: 50% Formamide, 5× SSPE, 5× Denhardt's, 0.5% SDS, 20 μg/mL ssDNA (For use of LNA-based probes, *see* **Note 2**).
9. X-ray film or Phosphor-imaging system.

3 Methods

3.1 Transfection

1. Hela cells are maintained in cell culture medium. Hela cells are efficiently transfected with Lipofectamine 3000 (Invitrogen) to deliver mirtron and myc-Ago2 encoding plasmids, as described above. Transfection and splicing efficiency is determined by visualizing GFP fluorescence.
2. Grow cells to 100% confluency, and change media 3 h before transfection with antibiotics free media.
3. Warm OptiMEM to room temperature. Make a mixture of OptiMEM, Plasmids, and Lipofectamine 3000 based on the scale of transfection in accordance with manufacturer recommendations. For this protocol, six well plates are typically used.
4. Incubate the mixture at room temperature for 20–25 min.
5. Add the transfection mixture to the cells dropwise, without touching the plate wall.
6. Swirl the dish gently and incubate for 2 days (37 °C, 5% CO_2).
7. Check transfection efficiency 2 days after transfection.

3.2 RNA Isolation

Following the two-day incubation, RNA is harvested. This could be done using the TRIzol method, following manufacturer protocols, or immunoprecipitation to isolate Ago-miRNA complexes.

1. TRIzol extraction (optional).
2. Add TRIzol directly to tissue culture plates. Cells should immediately become opaque, and easily dislodged by micropipetting. Remove cells and place in microcentrifuge tube.
3. Incubate TRIzol and cells for 10 min at room temperature.
4. Add 200 μL of chloroform to every 1 mL of TRIzol and incubate at room temperature for 5 min.
5. Separate phases at 4 °C for 15 min by spinning at 17,000 RCF.
6. Transfer the clear aqueous phase to an equal volume of isopropanol and incubate overnight at −20 °C.
7. Spin precipitations at 4 °C for 30 min, resuspend pellets in RNase-free water, and measure concentration via spectrophotometry.

3.3 Immunoprecipitation of miRNA-Ago Complexes (Optional)

1. Wash transfected cells with ice-cold PBS twice while in culture dish. Add lysis buffer to cells and incubate 20 min on ice. Cell layer should begin to detach from the bottom of the dish with gentle micropipetting. Transfer to microcentrifuge tube, mix well with pipette to eliminate clumps, and spin down at 17,000 RCF for 20 min at 4 °C to clarify lysate. Transfer supernatant to new tube.
2. Prepare beads for incubation with lysate. Add protein G magnetic beads to tubes (two tubes for each sample). Place the tubes on a magnetic stand and remove the supernatant. Wash tubes twice with PBS containing 0.02% tween-20. In PBS-tween add Anti-myc or control IgG antibodies.
3. Remove PBS tween from beads and add cell lysate from **step 1** of Subheading 3.3. Reserve 10% of the volume to be used as input control. Place on rocker for 2 h, 4 °C.
4. Wash beads with lysis buffer five times using magnetic stand to remove beads from solution.
5. Add equal volumes of 400 mM NaCl and acid phenol chloroform to beads. Aggressively vortex. This procedure should also be performed on the reserved 10% volume of lysate set aside for input control.
6. Separate phases at 4 °C for 15 min by spinning at 17,000 RCF.
7. Add clear aqueous phase to an equal volume of isopropanol and incubate overnight at −20 °C.
8. The following day spin precipitations at 4 °C for 30 min, resuspend pellets in RNA II loading buffer.

3.4 Northern Blotting

1. Sample Preparation. If immunoprecipitation was performed, samples are ready to load on the gel. If TRIzol extraction was used, 20 μg of RNA should be loaded in each lane.
2. Reprecipitate 20 μg of RNAs by adding the appropriate volume of RNA to 100% ethanol containing 300 mM sodium acetate.
3. Incubate the RNA at −20 °C for at least 1 h.
4. Spin precipitations at 4 °C for 30 min, resuspend pellets in RNA II loading buffer.
5. RNAs dissolved in RNA II loading buffer are denatured by incubating at 94 °C for 3 min and then immediately placed on ice.

3.5 Prepare and Run Gel Using a Vertical Gel Rig and the UreaGel System

1. Mix UreaGel diluent, buffer, and concentrate, then add APS and TEMED.
2. Quickly, using a large volume pipette pour gel mixture between glass plates, insert comb, and lay horizontally until gel is set.

3. Place gel on apparatus and fill buffer tanks with TBE. Pre-run the gel at 250 V for 30 min.
4. Using long gel loading tips, transfer RNA dissolved in RNA II loading buffer to the wells of the gel. Prior to loading samples, wells should be washed out to remove urea that has diffused into the space. Radio-labeled decade marker should also be loaded on the gel (*see* **Note 3** and **4**).
5. Run the gel at 250–350 V until the fastest moving, bromophenol blue dye band has migrated approximately 8 inch (*see* **Note 5**).

3.6 Blotting of RNAs onto Nylon Membranes

1. Soak filter paper in 0.5× TBE buffer and place onto wet-dry transfer rig.
2. Crack the gel using a separator wedge, and cut the region of the gel to be blotted. The lower dye band corresponds to ~10 nt RNAs. The gel should be cut approximately 1 inch. below the lower band in order to capture all the RNA molecules.
3. Place the gel onto the 0.5× TBE soaked filter paper.
4. Prepare the membrane by quickly rinsing in water, followed by briefly soaking in 0.5× TBE. Once the membrane sinks in the buffer, it is ready to be placed on top of the gel (*see* **Note 6**).
5. Place another presoaked piece of filter paper on top of the membrane.
6. Place the lid on the rig. The filter paper/gel/membrane should be sandwiched tightly between the electrodes.
7. Transfer the RNAs to the membrane for 30 min at 10 V at 4 °C.
8. Remove membrane from transfer apparatus, and air dry briefly on a paper towel.
9. Crosslink the RNAs to the membrane with UV exposure.
10. Bake the blot at 80 °C for 30 min.

3.7 Probing and Visualizing Membranes

1. Blots are placed into cylindrical hybridization tubes along with enough hybridization buffer to keep the blot wet.
2. Prehybridize blots for >1 h at 4–45 °C, in a rotating hybridization oven.
3. Prepare radio-labeled probes by mixing water, complementary DNA oligo, 10× PNK buffer, PNK, and gamma-P^{32}ATP 6000 μCi/mL.
4. Incubate the labeling reaction at 37 °C for >1 h.
5. Stop the reaction by adding EDTA.
6. Remove unincorporated radionucleotides using G-25 sephadex columns. Remove buffer from column by spinning at 400 RCF for 4 min. Transfer the probe-labeling reaction to the

column and spin again for 400 RCF, 4 min. The probe is in the column flowthrough.

7. Add the radio-labeled probe to the hybridization solution. Avoid touching the membrane with concentrated probe.
8. Incubate the membrane overnight at 40–45 °C in a rotating hybridization oven.

3.8 Wash and Expose the Blot

1. Wash the membrane twice with high stringency wash buffer at the hybridization temperature for 1 h.
2. Wash the membrane twice with low stringency wash buffer at the hybridization temperature for 1 h.
3. Remove membranes from tubes, and dab on filter paper. Do not over-dry the membrane. The objective is to remove excess liquid.
4. Wrap the membrane with cling wrap and expose to X-ray film or to phosphorimager screens (*see* **Note** 7).
5. Develop films or scan screens after 1 day.

3.9 Reprobing Membranes

1. Signal can be stripped from blots by boiling membranes in 0.1% SDS. Blots should be placed in boiling solution and left to cool for 3 h. This should be repeated once.
2. Following the stripping procedure, blots are placed into hybridization buffer and the procedure repeated starting at **step 2** of Subheading 3.4.
3. Blots can be subject to multiple reprobings before the integrity of the immobilized RNA is compromised.

4 Notes

1. The protocol described here relies on a radiolabel to visualize the interaction of probes with RNAs immobilized on blots. There are alternatives to this type of detection, which are untried with mirtrons [34]. These procedures involve an epitope-labeled probe such as biotin followed by antibody detection leading to visualization through chemiluminescence. For labs that want to avoid isotope, this could be a viable alternative [35].
2. This protocol calls for DNA oligonucleotide probes, but is also appropriate for use with LNA-based probes. These molecules offer greater specificity compared to DNA probes. To use this alternative chemistry, the hybridization buffer should be substituted with a formulation containing formamide [35, 36].
3. When removing combs from Urea Gels, the wells should be washed aggressively with water to prevent residual acrylamide from polymerizing and causing unevenness in the bottom of the well which will result in RNA migration artifacts.

4. Care should be taken to load the optimal amount of RNA ladder. The Decade Marker protocol calls for a single μL of gamma-P^{32}ATP 6000 μCi/mL, but this could lead to inappropriately intense bands that obscure signal if not properly diluted. When fresh ATP is used, the marker should be serially diluted in RNA II loading buffer. 1 μL of reaction should be diluted in 30 μL followed by a second dilution of 1 μL in 30 μL. One μL of the final dilution should be loaded on the gel [31].
5. Moderation of voltage used during PAGE separation of RNAs is recommended. This prevents bending of bands [31].
6. Avoid touching nylon membranes with gloves, and then only on the edges. This will avoid the appearance of blotting artifacts that might obscure signal [37].
7. Following the final wash step it is typical to observe very little signal emanating from the blot via Geiger counter. However, when signals are very weak longer exposure times may be required. Experimenters could extend exposure times to a week for the lowest signals [31].

References

1. Huang Q, Mao Z, Li S, Hu J, Zhu Y (2014) A non-radioactive method for small RNA detection by northern blotting. Rice 7(1):26
2. Westholm JO, Lai EC (2011) Mirtrons: microRNA biogenesis via splicing. Biochimie 93(11):1897–1904. https://doi.org/10.1016/j.biochi.2011.06.017
3. Fendereski M, Zia MF, Shafiee M, Safari F, Saneie MH, Tavassoli M (2017) MicroRNA-196a as a potential diagnostic biomarker for esophageal squamous cell carcinoma. Cancer Investig 35(2):78–84. https://doi.org/10.1080/07357907.2016.1254228
4. Meunier J, Lemoine F, Soumillon M, Liechti A, Weier M, Guschanski K, Hu H, Khaitovich P, Kaessmann H (2013) Birth and expression evolution of mammalian microRNA genes. Genome Res 23(1):34–45
5. Berezikov E (2011) Evolution of microRNA diversity and regulation in animals. Nat Rev Genet 12(12):846–860
6. Witten D, Tibshirani R, Gu SG, Fire A, Lui W-O (2010) Ultra-high throughput sequencing-based small RNA discovery and discrete statistical biomarker analysis in a collection of cervical tumours and matched controls. BMC Biol 8(1):58
7. Zhou J, Cheng Y, Yin M, Yang E, Gong W, Liu C, Zheng X, Deng K, Ren Z, Zhang Y (2015) Identification of novel miRNAs and miRNA expression profiling in wheat hybrid necrosis. PLoS One 10(2):e0117507
8. Lee Y, Ahn C, Han J, Choi H, Kim J, Yim J, Lee J, Provost P, Radmark O, Kim S, Kim VN (2003) The nuclear RNase III Drosha initiates microRNA processing. Nature 425(6956):415–419
9. Ketting RF, Fischer S, Bernstein E, Sijen T, Hannon GJ, Plasterk R (2001) Dicer functions in RNA interfence and in synthesis of small RNA involved in developmental timing in *C. elegans*. Genes Dev 15:2654–2659
10. Song J-J, Smith SK, Hannon GJ, Joshua-Tor L (2004) Crystal structure of Argonaute and its implications for RISC slicer activity. Science 305(5689):1434–1437
11. Babiarz JE, Ruby JG, Wang Y, Bartel DP, Blelloch R (2008) Mouse ES cells express endogenous shRNAs, siRNAs, and other microprocessor-independent, dicer-dependent small RNAs. Genes Dev 22(20):2773–2785
12. Ender C, Krek A, Friedländer MR, Beitzinger M, Weinmann L, Chen W, Pfeffer S, Rajewsky N, Meister G (2008) A human snoRNA with MicroRNA-like functions. Mol Cell 32(4): 519–528
13. Maute RL, Schneider C, Sumazin P, Holmes A, Califano A, Basso K, Dalla-Favera R (2013) tRNA-derived microRNA modulates proliferation and the DNA damage response and is

down-regulated in B cell lymphoma. Proc Natl Acad Sci U S A 110(4):1404–1409. https://doi.org/10.1073/pnas.1206761110

14. Yoshikawa M, Fujii YR (2016) Human ribosomal RNA-derived resident MicroRNAs as the transmitter of information upon the cytoplasmic cancer stress. Biomed Res Int 2016:7562085. https://doi.org/10.1155/2016/7562085
15. Ruby JG, Jan CH, Bartel DP (2007) Intronic microRNA precursors that bypass Drosha processing. Nature 448(7149):83–86
16. Macfarlane LA, Murphy PR (2010) MicroRNA: biogenesis, function and role in cancer. Curr Genomics 11(7):537–561. https://doi.org/10.2174/138920210793175895
17. Curtis HJ, Sibley CR, Wood MJ (2012) Mirtrons, an emerging class of atypical miRNA. Wiley Interdiscip Rev 3(5):617–632
18. Ladewig E, Okamura K, Flynt AS, Westholm JO, Lai EC (2012) Discovery of hundreds of mirtrons in mouse and human small RNA data. Genome Res 22(9):1634–1645. https://doi.org/10.1101/gr.133553.111
19. Westholm JO, Lai EC (2011) Mirtrons: microRNA biogenesis via splicing. Biochimie 93(11):1897–1904
20. Wen J, Ladewig E, Shenker S, Mohammed J, Lai EC (2015) Analysis of nearly one thousand mammalian Mirtrons reveals novel features of dicer substrates. PLoS Comput Biol 11(9):e1004441. https://doi.org/10.1371/journal.pcbi.1004441
21. Norbury CJ (2013) Cytoplasmic RNA: a case of the tail wagging the dog. Nat Rev Mol Cell Biol 14(10):643–653
22. Bortolamiol-Becet D, Hu F, Jee D, Wen J, Okamura K, Lin C-J, Ameres SL, Lai EC (2015) Selective suppression of the splicing-mediated microRNA pathway by the terminal uridyltransferase tailor. Mol Cell 59(2):217–228
23. Griffiths-Jones S, Grocock RJ, van Dongen S, Bateman A, Enright AJ (2006) miRBase: microRNA sequences, targets and gene nomenclature. Nucleic Acids Res 34(suppl_1):D140
24. Lee RC, Feinbaum RL, Ambros V (1993) The C. elegans heterochronic gene lin-4 encodes small RNAs with antisense complementarity to lin-14. Cell 75(5):843–854
25. Varkonyi-Gasic E, Wu R, Wood M, Walton EF, Hellens RP (2007) Protocol: a highly sensitive RT-PCR method for detection and quantification of microRNAs. Plant Methods 3(1):12
26. Sharbati-Tehrani S, Kutz-Lohroff B, Bergbauer R, Scholven J, Einspanier R (2008) miR-Q: a novel quantitative RT-PCR approach for the expression profiling of small RNA molecules such as miRNAs in a complex sample. BMC Mol Biol 9(1):34
27. Hwang H-W, Wentzel EA, Mendell JT (2009) Cell–cell contact globally activates microRNA biogenesis. Proc Natl Acad Sci 106(17):7016–7021
28. Rio DC (2014) Northern blots for small RNAs and microRNAs. Cold Spring Harb Protoc 2014(7):793–797. https://doi.org/10.1101/pdb.prot080838
29. Xu H, Wang P, Fu Y, Zheng Y, Tang Q, Si L, You J, Zhang Z, Zhu Y, Zhou L (2010) Length of the ORF, position of the first AUG and the Kozak motif are important factors in potential dual-coding transcripts. Cell Res 20(4):445–457
30. Taji F, Kouchesfahani HM, Sheikholeslami F, Romani B, Baesi K, Vahabpour R, Edalati M, Teimoori-Toolabi L, Jazaeri EO, Abdoli A (2017) Autophagy induction reduces telomerase activity in HeLa cells. Mech Ageing Dev 163:40–45. https://doi.org/10.1016/j.mad.2016.12.011
31. Pall GS, Hamilton AJ (2008) Improved northern blot method for enhanced detection of small RNA. Nat Protoc 3(6):1077–1084. https://doi.org/10.1038/nprot.2008.67
32. Josefsen K, Nielsen H (2011) Northern blotting analysis. Methods Mol Biol 703:87–105. https://doi.org/10.1007/978-1-59745-248-9_7
33. Parmar S, Buchner P, Hawkesford M (2007) Leaf developmental stage affects sulfate depletion and specific sulfate transporter expression during sulfur deprivation in Brassica napus L. Plant Biol 9(05):647–653
34. Kloosterman WP, Wienholds E, Ketting RF, Plasterk RHA (2004) Substrate requirements for let-7 function in the developing zebrafish embryo. Nucl Acids Res 32(21):6284–6291
35. Hermeking H, Lengauer C, Polyak K, He TC, Zhang L, Thiagalingam S, Kinzler KW, Vogelstein B (1997) 14-3-3sigma is a p53-regulated inhibitor of G2/M progression. Mol Cell 1(1):3–11
36. Kaghad M, Bonnet H, Yang A, Creancier L, Biscan JC, Valent A, Minty A, Chalon P, Lelias JM, Dumont X, Ferrara P, McKeon F, Caput D (1997) Monoallelically expressed gene related to p53 at 1p36, a region frequently deleted in neuroblastoma and other human cancers. Cell 90(4):809–819
37. Brown T, Mackey K, Du T (2004) Analysis of RNA by northern and slot blot hybridization. Curr Protoc Mol Biol Chapter 4:Unit 4.9. Doi:https://doi.org/10.1002/0471142727.mb0409s67

Chapter 17

Detecting Agotrons in Ago CLIPseq Data

Thomas B. Hansen

Abstract

Timely and accurate regulation of gene expression is essential to all organisms. MicroRNAs (miRNAs) are a well-characterized and important class of gene expression regulators. We recently identified a novel class of gene regulators, the agotrons. Agotrons derive from short introns and associate with Argonaute (Ago) proteins much similar to miRNAs. However, agotrons completely bypass the conventional miRNA biogenesis pathway and thus exist as full-length introns, which disobey the classical rules on Ago-substrate requirements. As a class, agotrons are conserved in mammals, and despite the non-canonical biogenesis pathway, agotrons maintain the ability to deregulate mRNAs with seed-matches in the 3′UTR. While several pipelines exist for the detection of miRNAs, no bioinformatics toolkit has yet been developed to specifically identify agotrons. Here, we describe a simple approach, termed agotron_detector (https://github.com/ncrnalab/agotron_detector), to identify and quantify agotrons in Ago CLIPseq datasets. Hopefully, this allows researchers worldwide to characterize agotrons in more detail and to reveal the biological relevance of these fascinating RNA species.

Key words Agotron, Mirtron, miRNA, Argonaute, Ago, CLIP, RNA sequencing, Small RNA profiling, Gene regulation

1 Introduction

The existence of miRNAs has been known since the millennium, and eminent research since then has disclosed the steps of biogenesis and the effects of miRNA on mRNA stability and translational activity (reviewed in [1, 2]), although the mechanistic details are still debated. Moreover, several subclasses of miRNAs that diverge from the consensus route of biogenesis have emerged [3]: First the discovery of mirtrons, that bypass Microprocessor cleavage but instead produce the pre-miRNAs hairpin by splicing [4–6]. Other examples comprise miR-451 that requires Ago2-mediated slicing instead of Dicer cleavage for maturation [7, 8], and 5′ capped pre-miRNAs, so-called endo-shRNA, that bypass the Microprocessor but require Dicer for loading of the 3p strand selectively [9]. During the last decade, much effort has been put into disclosing the miRNAome, i.e., the total content of regulatory small RNAs within human cells

Ulf Andersson Vang Ørom (ed.), *miRNA Biogenesis: Methods and Protocols*, Methods in Molecular Biology, vol. 1823, https://doi.org/10.1007/978-1-4939-8624-8_17, © Springer Science+Business Media, LLC, part of Springer Nature 2018

and tissues, and the effect on gene expression. In addition, biochemical analyses of Ago-associated RNA have been conducted thoroughly over the years with increasing depth to characterize the content, origin, and molecular signature of Ago-loaded entities and targets, and as of today many pipelines exist to scrutinize and annotate miRNAs in high-throughput small RNA sequencing [10]. Recently, we identified a new class of Ago-associates species, the agotrons, that seemingly associate with Ago proteins as full-length but short introns, thereby bypassing both Microprocessor and Dicer processing [11]. These species are in the 50–150 nucleotide (nt) size range and thus typically escape detection in classical small RNA sequencing. Instead, agotrons are evident in Ago CLIPseq, e.g., HITS-CLIP, PAR-CLIP, and iCLIP, experiments. Apart from mapping to short introns, the agotrons exhibit a longer CLIPseq read-length compared to conventional miRNAs/mirtrons. Moreover, most agotron-derived reads align to the very 5′end of introns. Despite their relatively long mature length, agotrons have the potential to act as gene regulators similar to miRNAs. However, it is currently undisclosed how agotrons associate with Ago proteins, and what the true biological relevance of this interaction is [12].

Here, we provide a simple toolkit, coined *agotron_detector*, to identify agotrons in Ago CLIPseq data (schematically outlined in Fig. 1a). Briefly, the approach intersects genome-mapped reads with annotated short introns. Then, the quantity of reads, distribution of read-lengths, and the 5′ heterogeneity for each intron are analyzed and outputted. Apart from established third-party algorithms, the toolkit comprises three scripts that retrieve (*UCSC_intron_retriever.py*, *see* Subheading 3.5), analyze (*analyzer.py*, *see* Subheading 3.6) and annotate short introns (*annotater.R*, *see* Subheading 3.7), respectively, to identify and quantify agotron expression.

2 Materials

2.1 Required Software Packages

The toolkit described below requires the following third-party installations (Table 1). The specified releases are not necessarily mandatory, but this was the exact setup used for testing.

2.2 Genome Assemblies

To analyze high-throughput sequencing datasets, a reference genome is usually required to determine the genomic origin of reads in the dataset. Thus, prior to dataset analyses, the reference genome must be retrieved and indexed (*see* Subheading 3.1 for genome preparation).

Fig. 1 (continued) which was based on other Ago CLIPseq datasets. The numbers in parenthesis denote the top 10-expressed agotrons in GSE78059. *P*-value is determined by chi-square test. (**f**) The duration of each subsection running the *all_commands.sh* script (available on GitHub) on our server, from downloading genomes and datasets to agotron detection and quantification, see text for detailed description of each section

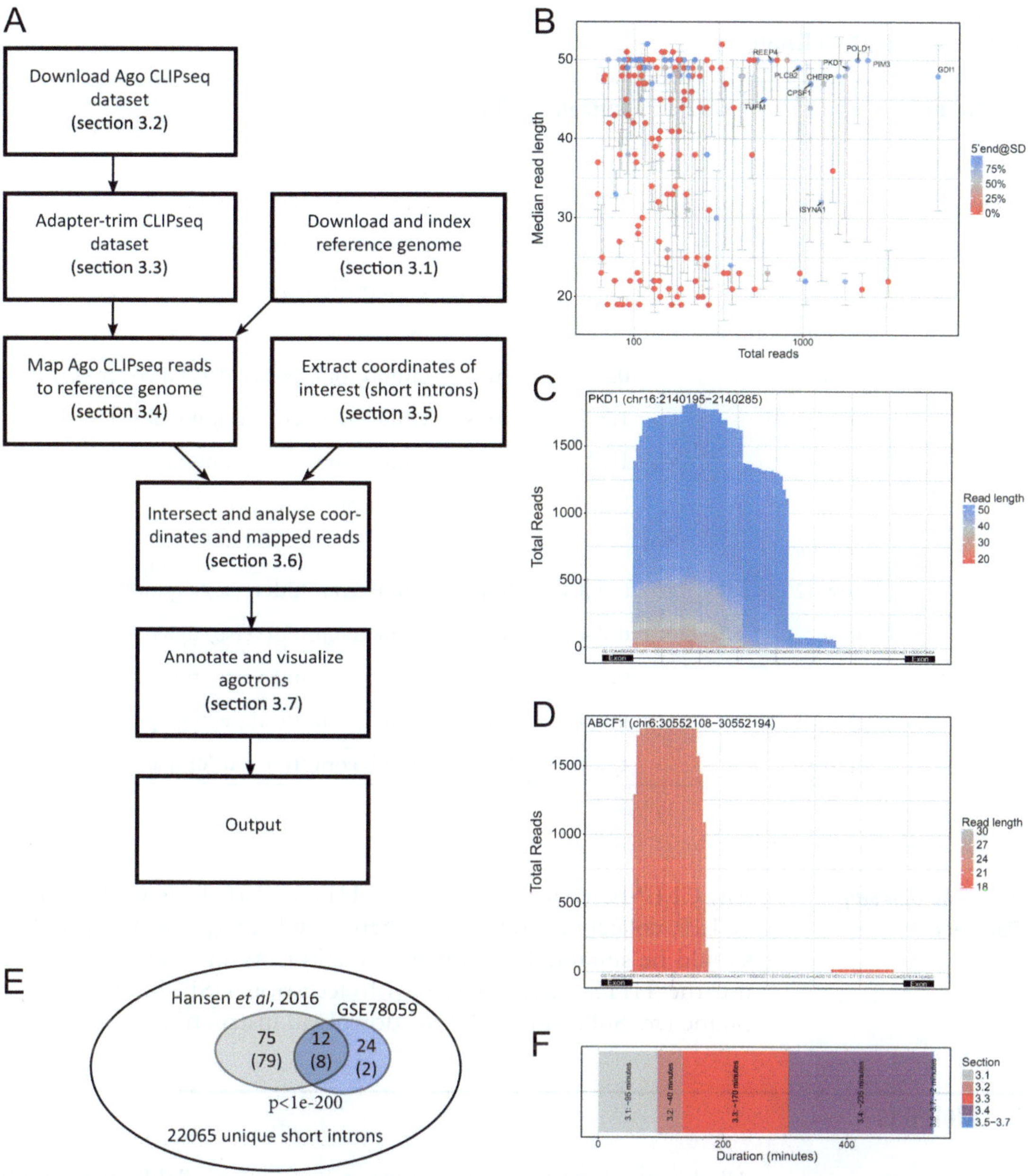

Fig. 1 The agotron_detector toolkit. (**a**) Schematic pipeline for the described toolkit, see text for more detail. (**b**) Scatterplot showing the median read-length as a function of total read count in all RefSeq-annotated short introns with at least 5 RPMM in at least two Ago CLIPseq datasets (from GSE78059). The error bars reflect the 5th and 95th percentile, respectively. The individual introns are color-coded by the fraction of 5′ends within 1 nucleotide from the 5′ splice site (SD, splice donor). The top 10-expressed agotrons (median read-length > 30 nts and > 70% 5′ends within 1 nt from SD) are labeled with host-gene names. (**c-d**) Density plot of highly conserved and abundant agotron derived from *PKD1* (C), and highly abundant mirtron (miR-877) derived from *ABCF1* (D) with the respective intron coordinates in the upper left corners. Reads are color-coded by the read-length as denoted in the legend, and the exon-intron structure is schematically depicted below the plot. (**e**) Out of a total of 22,065 unique short introns annotated by RefSeq, the Venn diagram shows predicted agotrons from the GSE78059 dataset used here (n = 36), and the agotrons annotated by Hansen et al., 2016 (n = 87)

Table 1
Required 3rd party software

Software		Version	URLs
trim_galore		0.4.1	hhttps://github.com/FelixKrueger/TrimGalore
bowtie2		2.2.8	https://github.com/BenLangmead/bowtie2
samtools		1.3	https://github.com/samtools/
python		2.7	https://www.python.org/
	Non-standard modules:		
	pysam	0.11.1	https://github.com/pysam-developers/pysam
	MySQLdb	1.2.5	https://github.com/arnaudsj/mysql-python
	numpy	1.10.1	https://github.com/numpy/numpy
R		3.2.5	https://www.r-project.org/
	Packages:		
	ggplot2	0.9.3.1	https://github.com/tidyverse/ggplot2
	dplyr	0.7.4	https://github.com/tidyverse/dplyr
	tidyr	0.7.2	https://github.com/tidyverse/tidyr
	ggrepel	0.7.0	https://github.com/slowkow/ggrepel
	optparse	1.4.4	https://github.com/trevorld/optparse

2.3 Ago CLIPseq Datasets

Any CLIP-based experiment (e.g., HITS-CLIP, PAR-CLIP, iCLIP, eCLIP, collectively referred to here as CLIPseq) on Ago proteins should be suitable for agotron detection. As an example, we here use the HITS-CLIP dataset with accession GSE78059 available online (*see* Subheading 3.2 for download instruction).

3 Methods

All the developed scripts presented here are available at GitHub (https://github.com/ncrnalab/agotron_detector). Furthermore, to avoid potential issues copy-pasting from a unicode formatted document, all the prompt commands shown below are collected into a complete bash-script, *all_commands.sh*, also available at GitHub. Overall, the following sections will demonstrate agotron analysis on Ago CLIPseq (accession GSE78059) all the way from downloading the reference genome (here using the hg19 assembly), mapping the sequencing data to the genome, and finally detecting agotrons.

3.1 Preparing Reference Genome

A reference genome is obviously required to interrogate the genome origin of reads in a high-throughput sequencing dataset. Downloading and preparing genomes are slightly time-consuming commands but only required once per genome.

1. To download the hg19 assembly from UCSC, execute the following command:

```
wget http://hgdownload.cse.ucsc.edu/
goldenPath/hg19/bigZips/chromFa.tar.gz
```

2. Unpack the genome:

```
tar -zxvf chromFa.tar.gz
```

3. Combine all fasta files in one:

```
cat *.fa > hg19.fa
```

4. Create fasta file index using samtools:

```
samtools faidx hg19.fa
```

5. Create bowtie2 index for mapping:

```
bowtie2-build hg19.fa hg19
```

3.2 Downloading Dataset

There are many ways to download datasets from online repositories. Here, we use a recently published Ago CLIPseq data [13], accession id GSE78059. The dataset was found at the European Nucleotide Archive (ENA, https://www.ebi.ac.uk/ena), and from here a list of fastq URLs was copy-pasted into a bash-script array. Subsequently, iterating through the array and invoking wget for each URL provides a simple method for downloading the fastq files:

6. Create a bash-script, e.g., entitled download.sh. For GSE78059, the script will look as follows:

```
#!/bin/bash
array=( \
ftp.sra.ebi.ac.uk/vol1/fastq/SRR317/008/
 SRR3177718/SRR3177718.fastq.gz \
ftp.sra.ebi.ac.uk/vol1/fastq/SRR317/009/
 SRR3177719/SRR3177719.fastq.gz \
ftp.sra.ebi.ac.uk/vol1/fastq/SRR317/000/
 SRR3177720/SRR3177720.fastq.gz \
ftp.sra.ebi.ac.uk/vol1/fastq/SRR317/001/
 SRR3177721/SRR3177721.fastq.gz \
ftp.sra.ebi.ac.uk/vol1/fastq/SRR317/002/
 SRR3177722/SRR3177722.fastq.gz \
ftp.sra.ebi.ac.uk/vol1/fastq/SRR317/003/
 SRR3177723/SRR3177723.fastq.gz)
for i in "${array[@]}"
do
      wget $i
done
```

7. Execute the script with the following command:

```
sh download.sh
```

3.3 Adaptor Trimming of RNA Sequencing Dataset

Typically, in a small RNA sequencing dataset such as CLIPseq, a substantial fraction of reads contains part of the 3′ adapter sequence. In addition, the 5′ end of reads often contain barcodes and unique molecular identifiers (UMIs), and importantly, prior to mapping, these library-specific sequences should be removed.

8. Trim 3′ adaptors off the raw fastq files with trim_galore using the following command:

```
trim_galore -A TCAGTCACTTCCAGC -length 18
*.fastq.gz
```

Here, the -A TCAGTCACTTCCAGC argument denoted the adaptor sequence, and -length 18 filters out reads below 18 nucleotides in length after adapter removal. Trim_galore outputs the processed reads in a file with postfix "_trimmed.fq.gz."

3.4 Mapping Reads to Genome

Many aligning algorithms exist for mapping nucleotide sequences to a reference genome, and the pros and cons of different approaches will not be discussed. We, however, expect that most aligners will be compatible with the described toolkit. Here, we use bowtie2 to assign a coordinate and a mapping quality score to each read in the fastq file. There are a number of options to tweak and modulate the performance of bowtie2, but we use the default setting except for enabling local alignments (set by the --local flag, see below). This permits soft-clipping and allows mapping of "tailed" reads with non-templated nucleotides often seen for miRNAs and mirtrons [14–16].

The following commands map the reads from a single adaptor-trimmed fastq file (*SRR3177718_trimmed.fq.gz*), then sorts the output and creates an index for fast coordinate queries required for subsequent analyses:

9. Map the reads to reference genome using bowtie2, e.g., for SRR3177718:

```
bowtie2 -q --local -x hg19 -U SRR3177718_
trimmed.fq.gz > SRR3177718.bam
```

The –q flag denotes that the reads are fastq-formatted, –x hg19 refers to the bowtie2-index (*see* Subheading 3.1, command 5), and the –U argument designates the path to the unpaired fastq file (in this case, *SRR3177718_trimmed.fq.gz*).

10. Sort the mapped reads with samtools:

```
samtools sort < SRR3177718.bam >
SRR3177718.sort.bam
```

11. Or alternatively, pipe the above two commands (9 and 10):

```
bowtie2 -q --local -x hg19 -U SRR3177718_
trimmed.fq.gz | samtools sort - >
SRR3177718.sort.bam
```

12. Index the bam-file using the following command:

```
samtools index SRR3177718.sort.bam
```

3.5 Getting Loci for Analysis

To retrieve gene annotation of an organism of interest, we take advantage of the openly available UCSC mySQL database. Using the MySQLdb API for python, a database containing genome annotations is easily queried and analyzed for each genome assembly (here we use "hg19"). As mentioned above, agotrons are by definition derived from introns, presumably short introns specifically. To extract short intron, the simple python script, *UCSC_intron_retriever.py* (*see* Table 2 for usage), extracts all short introns based on the RefSeq annotation available at the UCSC MySQL database:

13. Retrieve all RefSeq-annotated short introns:

```
python UCSC_intron_retriever.py > loci.bed
```

Instead of analyzing short intron in the attempt to detect and evaluate agotrons, it is also possible to retrieve pre-miRNA coordinates using the alternative script, *UCSC_mirna_retriever.py*, and then perform all downstream analyses on annotated miRNAs loci. This provides a simple and fast approach to quantify pre-miRNAs in Ago-CLIP dataset (or small RNAseq in general) and obtain read density across the pre-miRNA sequence (*see* Subheading 3.7). In any case, the output from the *retriever*-scripts is a conventional 6-column bed file, and this is replaceable by any bed-formatted coordinates of interest to intersect with bam-files as described below.

3.6 Intersecting Loci with Bam

From the bam-files (Subheading 3.4), the mapped reads derived from the short introns (*see* Subheading 3.5) are extracted and analyzed using the python script, *analyzer.py* (Table 3):

Table 2
Options and arguments for *UCSC_intron_retriever.py*

Usage: **UCSC_intron_retriever.py** [ARGUMENTS]	
Optional arguments:	
-db <string>	MySQL database (default='hg19')
-table <string>	MySQL table (default='refGene')
-max <int>	Maximum intron length (default=150)
-min <int>	Minimum intron length (default=50)

Table 3
Options and arguments for *analyzer.py*

Usage: **analyzer.py** [ARGUMENTS] < [INPUT] > [OUTPUT]	
Required arguments:	
-g <file>	Path to reference genome fastafile, must be indexed with samtools faidx
Optional arguments:	
-f <files...>	Input bam-files (default=*.bam)
-c <file>	Filename for coverage output (if empty, no coverage file is produced) (default='coverage.txt')
-tr <int>	RPMM (reads per mapped million) expression threshold for output (default = 5)
-ts <int or 'all'>	How many samples to meet RPMM expression threshold. For all samples, type 'all' (default = 2)
-m <float>	Tolerance for reads mapping partly outside locus (default=0.1, e.g., 10% of the reads is allowed to map outside locus)
-q <int>	Threshold for mapping quality (default=13)
-a <int>	Add <int> flanking nucleotides (up and downstream) to the loci sequence output (default=10)

The loci of introns (here short intron) are provided in the standard input and for each locus the intersecting reads are obtained from all the bam-files provided. The only required argument is the path for the genome fasta file -g. The script has two output, a bed file (standard out) providing information on read-length, 5′end homogeneity, and position as well as expression data for each bam-file, in case the locus of interest meets the expression threshold (−tr argument) in at least a subset of samples (−ts argument). In this step, agotrons are not distinguished from mirtrons and other loci with sufficient reads. The script also outputs read-coverage in a distinct file (set by the -c argument). If set, this provides the information used to generate density plots (Fig. 1c, d, described further in Subheading 3.7).

14. To intersect the retrieved loci with bowtie2-mapped bam-files:

```
python analyzer.py -g hg19.fa < loci.bed >
output.bed
```

3.7 Annotating agotrons

The above analysis (in Subheading 3.4) outputs 177 loci with at least 5 reads per mapped million (RPMM) found in least two datasets (the default settings). This, as mentioned, contains agotrons, mirtrons and, presumably, non-specific background noise. To annotate agotrons in the obtained loci, the R-script, *annotater.R*,

Table 4
Options and arguments for *annotater.R*

Usage: **annotater.R** [ARGUMENTS] < [INPUT]	
Optional arguments:	
-c <file>	Input coverage file (the -c output from analyzer.py) (default='coverage.txt')
-p <string>	Prefix used in output files (default='agotron')
Agotron definition:	
-m <int>	Threshold for median read length (default=30)
-h <float>	Minimum fraction of reads with distance (−d) between 5'end of read and 5'end of locus (default=0.7)
-d <int>	Maximum distance allowed from predominant 5'end of reads to 5'end of locus (default=1)

is tailored to filter out agotrons from other CLIPseq containing short introns (Table 4).

Currently, agotrons have not been thoroughly characterized and thus the defining features are still elusive. However, agotrons are by definition non-processed short introns and therefore significantly longer than dicer-processed RNA species. Consequently, the read-lengths obtained in Ago CLIPseq experiments are typically above 30 nts in length (determined by the –m setting) and not ~22 nts as expected for mirtrons and miRNAs. Moreover, the 5′end of agotrons are very homogeneous (−h) as seen for most if not all Ago-associated RNAs with functional relevance, and consistent with unprocessed introns, the 5′end of agotrons furthermore aligns with the 5′ splice site (splice donor, SD) or occasionally, as also observed for mirtrons [16], with 1 nt decapitation of the 5′ guanine (the –d setting).

15. To identify and annotate agotrons in the analyzed loci (using the output.bed from above as input):

```
Rscript annotater.R < output.bed
```

16. Collectively, extracting the loci of interest (13), intersecting with bowtie2-mapped reads (14), and filtering for agotrons (15) can be performed with the following pipe:

```
python UCSC_intron_retriever.py | py-
thon analyzer.py -g hg19.fa | Rscript
annotater.R
```

As a result, the pipe outputs a bed-like table containing all the filtered agotrons (entitled *[prefix]_filtered.bed*) containing in this case 36 putative agotron species with the associating RPMM and raw read counts obtained in the individual bam-files (exemplified with top5-expressed agotrons in Table 5). Moreover, a median

Table 5
The top5-expressed agotrons

						RPMM					
chr	Start	End	Name	Score	Strand	SRR3177718	SRR3177719	SRR3177720	SRR3177721	SRR3177722	SRR3177723
chrX	153,670,775	153,670,866	GDI1	6432	+	25.18	43.22	36.54	122.62	41.52	183.52
chr22	50,356,513	50,356,587	PIM3	2424	+	0.54	5.76	3.23	58.55	35.36	77.21
chr19	50,909,579	50,909,663	POLD1	2110	+	32.72	19.52	8.81	13.00	11.94	58.72
chr16	2,140,195	2,140,285	PKD1	1824	–	0.23	17.05	17.62	33.84	20.45	35.46
chr19	16,631,076	16,631,174	CHERP	1625	–	17.64	13.17	6.21	21.57	18.32	38.21

Output slightly modified from the *[prefix] _filtered.txt* produced by *annotater.R* read-length by read count scatterplot depicting all the short introns passing the expression filtering is produced (Fig. 1b, entitled *[prefix] _scatter.pdf*) color-coded by the fraction of reads with 5′ends within one nucleotide from the SD). Also, for a subset of loci, the density of reads across the short introns with 10 nt flanking sequence on each side (the –a option in Subheading 3.6) is visualized and color-coded by median read-length showing a clear demarcation between agotrons and mirtrons (Fig. 1c, d, respectively, entitled *[prefix]_coverage.pdf*).

4 Discussion

Here, based on one Ago CLIPseq dataset (GSE78059), we identify 36 agotrons using the default cutoff. One-third of these agotrons (12 of 36) were also identified previously [11] (Fig. 1e). Currently, agotrons are still largely uncharacterized territory, and presumably, hundreds of agotrons have yet to be identified and annotated. However, 8 of the top10-expressed agotrons in GSE78059 have been reported previously, suggesting that there may be a restricted subset of widely abundant and commonly expressed agotrons. Apart from the genome preparation and the retrieval of Ago CLIPseq datasets (Subheadings 3.1–3.3), mapping of reads in our setup took approximately 40 min per sample and consumed less than 8 g of RAM. Subsequently, analyzing the mapped reads (command 16) finished within 2 min, and therefore the described pipeline provides a toolkit that simply and quickly annotates and quantifies agotrons in Ago CLIPseq datasets (Fig. 1f).

Conclusively, we hope that this toolkit will constitute a platform for agotron identification and annotation, and that other research laboratories will benefit and build upon this pipeline in the coming years.

Acknowledgments

The computational analyses were conducted on the GenomeDK HPC cluster, Aarhus University. This work was supported by the Novo Nordisk Foundation (NNF16OC0019874).

References

1. Bartel DP (2009) MicroRNAs: target recognition and regulatory functions. Cell 136:215–233
2. Jonas S, Izaurralde E (2015) Towards a molecular understanding of microRNA-mediated gene silencing. Nat Rev 16:421–433
3. Daugaard I, Hansen TB (2017) Biogenesis and function of ago-associated RNAs. Trends Genet 33:208–219
4. Berezikov E, Chung WJ, Willis J, Cuppen E, Lai EC (2007) Mammalian mirtron genes. Mol Cell 28:328–336
5. Okamura K, Hagen JW, Duan H, Tyler DM, Lai EC (2007) The mirtron pathway generates microRNA-class regulatory RNAs in drosophila. Cell 130:89–100
6. Ruby JG, Jan CH, Bartel DP (2007) Intronic microRNA precursors that bypass Drosha processing. Nature 448:83–86
7. Cheloufi S, Santos COD, Chong MM, Hannon GJ (2010) A dicer-independent miRNA biogenesis pathway that requires ago catalysis. Nature 465:584–589
8. Cifuentes D, Xue H, Taylor DW, Patnode H, Mishima Y, Cheloufi S, Ma E, Mane S, Hannon GJ, Lawson ND et al (2010) A novel miRNA processing pathway independent of dicer requires Argonaute2 catalytic activity. Science 328:1694–1698
9. Xie M, Li M, Vilborg A, Lee N, Shu MD, Yartseva V, Sestan N, Steitz JA (2013) Mammalian 5′-capped microRNA precursors that generate a single microRNA. Cell 155:1568–1580
10. Kang W, Friedländer MR (2015) Computational prediction of miRNA genes from small RNA sequencing data. Front Bioeng Biotechnol 3:7
11. Hansen TB, Veno MT, Jensen TI, Schaefer A, Damgaard CK, Kjems J (2016) Argonaute-associated short introns are a novel class of gene regulators. Nat Commun 7:11538
12. Stagsted LVW, Daugaard I, Hansen TB (2017) The agotrons: gene regulators or Argonaute protectors? BioEssays 39:1600239
13. Gillen AE, Yamamoto TM, Kline E, Hesselberth JR, Kabos P (2016) Improvements to the HITS-CLIP protocol eliminate widespread mispriming artifacts. BMC Genomics 17:338
14. Ameres SL, Horwich MD, Hung JH, Xu J, Ghildiyal M, Weng Z, Zamore PD (2010) Target RNA-directed trimming and tailing of small silencing RNAs. Science 328:1534–1539
15. Hansen TB, Veno MT, Kjems J, Damgaard CK (2014) miRdentify: high stringency miRNA predictor identifies several novel animal miRNAs. Nucleic Acids Res 42:e124
16. Wen J, Ladewig E, Shenker S, Mohammed J, Lai EC (2015) Analysis of nearly one thousand mammalian mirtrons reveals novel features of dicer substrates. PLoS Comput Biol 11:e1004441

Index

Ulf Andersson Vang Ørom (ed.), *miRNA Biogenesis: Methods and Protocols*, Methods in Molecular Biology, vol. 1823,
https://doi.org/10.1007/978-1-4939-8624-8, © Springer Science+Business Media, LLC, part of Springer Nature 2018

N

U

V

W

MIX
Papier aus verantwortungsvollen Quellen
Paper from responsible sources
FSC® C105338

If you have any concerns about our products,
you can contact us on
ProductSafety@springernature.com

In case Publisher is established outside the EU,
the EU authorized representative is:
Springer Nature Customer Service Center GmbH
Europaplatz 3, 69115 Heidelberg, Germany

Printed by Libri Plureos GmbH
in Hamburg, Germany